AVAILABILITY NOTICE

Availability of Reference Materials Cited in NRC Publications

Most documents cited in NRC publications will be available from one of the following sources:

1. The NRC Public Document Room, 2120 L Street, NW, Lower Level, Washington, DC 20555

2. The Superintendent of Documents, U.S. Government Printing Office, P.O. Box 37082, Washington, DC 20013-7082

3. The National Technical Information Service, Springfield, VA 22161

Although the listing that follows represents the majority of documents cited in NRC publications, it is not intended to be exhaustive.

Referenced documents available for inspection and copying for a fee from the NRC Public Document Room include NRC correspondence and internal NRC memoranda; NRC Office of Inspection and Enforcement bulletins, circulars, information notices, inspection and investigation notices; Licensee Event Reports; vendor reports and correspondence; Commission papers; and applicant and licensee documents and correspondence.

The following documents in the NUREG series are available for purchase from the GPO Sales Program: formal NRC staff and contractor reports, NRC-sponsored conference proceedings, and NRC booklets and brochures. Also available are Regulatory Guides, NRC regulations in the *Code of Federal Regulations*, and *Nuclear Regulatory Commission Issuances*.

Documents available from the National Technical Information Service include NUREG series reports and technical reports prepared by other federal agencies and reports prepared by the Atomic Energy Commission, forerunner agency to the Nuclear Regulatory Commission.

Documents available from public and special technical libraries include all open literature items, such as books, journal and periodical articles, and transactions. *Federal Register* notices, federal and state legislation, and congressional reports can usually be obtained from these libraries.

Documents such as theses, dissertations, foreign reports and translations, and non-NRC conference proceedings are available for purchase from the organization sponsoring the publication cited.

Single copies of NRC draft reports are available free, to the extent of supply, upon written request to the Office of Information Resources Management, Distribution Section, U.S. Nuclear Regulatory Commission, Washington, DC 20555.

Copies of industry codes and standards used in a substantive manner in the NRC regulatory process are maintained at the NRC Library, 7920 Norfolk Avenue, Bethesda, Maryland, and are available there for reference use by the public. Codes and standards are usually copyrighted and may be purchased from the originating organization or, if they are American National Standards, from the American National Standards Institute, 1430 Broadway, New York, NY 10018.

DISCLAIMER NOTICE

This report was prepared as an account of work sponsored by an agency of the United States Government. Neither the United States Government nor any agency thereof, or any of their employees, makes any warranty, expressed or implied, or assumes any legal liability of responsibility for any third party's use, or the results of such use, of any information, apparatus, product or process disclosed in this report, or represents that its use by such third party would not infringe privately owned rights.

NUREG/CR–5383
RV

Effect of Aging on Response Time of Nuclear Plant Pressure Sensors

Manuscript Completed: May 1989
Date Published: June 1989

Prepared by
H. M. Hashemian, K. M. Petersen,
R. E. Fain, J. J. Gingrich

Analysis and Measurement Services Corporation
AMS 9111 Cross Park Drive, NW
Knoxville, TN 37923–4599

Prepared for
Division of Engineering
Office of Nuclear Regulatory Research
U.S. Nuclear Regulatory Commission
Washington, DC 20555
NRC FIN D2503

ISBN-13: 978-1499140118; ISBN-10: 1499140118

This manuscript has been authored by a contractor of the U.S. Government under Grant No. D 2503. Accordingly, the U.S. Government has a nonexclusive, royalty-free license to publish or reproduce the published form of this contribution, or allow others to do so, for U.S. Government purposes.

ABSTRACT

A research program was initiated to study the effects of normal aging on the dynamic performance of safety-related pressure transmitters in nuclear power plants. The project began with an experimental assessment of the conventional and new testing methods for measurement of response time of pressure transmitters. This was followed by developing a laboratory set up and performing initial tests to study the aging characteristics of representative transmitters of the type used in nuclear power plants.

The project also included a search of the LER data base for pressure sensing system problems and a review of the Regulatory Guide 1.118 and the industry standards on performance testing of pressure transmitters. The following conclusions have been reached:

- Five reasonably effective methods are available for response time testing of pressure transmitters in nuclear power plants. These methods are referred to as step test, ramp test, frequency test, noise analysis, and power interrupt test. Two of the five methods (noise analysis and power interrupt test) have the advantage of providing on-line measurement capability at normal operating conditions.

- The consequences of aging at simulated plant conditions were calibration shifts and response time degradation, with the former being the more pronounced problem.

- The LER data base contains 1,325 cases of reported problems with pressure sensing systems over a nine year period. Potential age-related cases account for 38 percent of the reported problems in this period. A notable number of LERs reported problems with sensing lines. These problems include sensing line blockages, freezing, and void.

- Regulatory Guide 1.118, IEEE Standard 338, and ISA Standard 67.06 can benefit from minor revisions to account for recent advances in performance testing technologies and from new information that has become available since these documents were prepared.

The work reported herein was a feasibility study performed over a six-month period. As such, the results and conclusions presented herein are preliminary.

TABLE OF CONTENTS

LIST OF FIGURES

LIST OF TABLES

ACKNOWLEDGEMENTS

The cooperation of a number of organizations were very helpful during this research.

Oak Ridge National Laboratory, Sandia National Laboratories, Arkansas Power and Light Company, and Tobar Incorporated provided nuclear-type pressure transmitters for the laboratory tests. The contributions of these organizations are gratefully acknowledged.

1. INTRODUCTION

This report presents the results of a preliminary study of the dynamic performance of pressure transmitters. The focus of the project has been on pressure, level, and flow transmitters (hereafter referred to as pressure transmitters) that are used for safety-related measurements in nuclear power plants. The words pressure transmitter and pressure sensor are used in this report interchangeably.

A six-month study has been completed covering the following areas:

- Assessment of Response Time Testing Methods. Five methods are available for response time testing of pressure transmitters. These are called ramp test, step test, frequency test, noise analysis and power interrupt (PI) test. An experimental assessment of these methods was performed. The assessment involved laboratory testing of more than twenty pressure transmitters with all five methods. The work concluded that the methods are equally effective, but vary widely in the degree of difficulty in implementation in nuclear power plants. Two of the five methods (noise analysis and power interrupt test) can be performed remotely on installed transmitters while the plant is at normal operating conditions.

- Aging Study. Laboratory research was initiated and preliminary aging results were obtained. The work involved response time testing and calibration checks of a number of transmitters after exposure to heat, humidity, vibration, pressure cycling, and overpressurization conditions. The effect of these conditions was response time degradation and calibration shifts, with the latter being the more pronounced problem.

- Review of Historical Data. The Licensee Event Report (LER) data base was searched for pressure transmitter problems. The search covered the period beginning in 1980 until October 1988. The search revealed 1,325 reports of failure or degradation of pressure sensing systems, 498 cases of which were considered as potential age-related problems. The number of reported problems with pressure transmitters dropped by a factor of about two after 1984, when the LER reporting requirements were changed.

- Review of Regulatory Guide and Industry Standards. The provisions of Regulatory Guide 1.118 as related to performance testing of safety system sensors were reviewed along with IEEE Standard 338 and ISA Standard 67.06. This review concluded that the Regulatory Guide and both standards should be revised to reflect the current practices.

- <u>Review of Related Studies.</u> All of the related experimental work on aging of pressure transmitters has concentrated on the effects of aging on static performance of transmitters as opposed to the dynamic performance reported herein. The related studies have concluded that aging affects the performance of pressure transmitters and that temperature is the dominant stressor. Most of the studies on performance of nuclear plant pressure transmitters have been sponsored by the U.S. Nuclear Regulatory Commission (NRC). The only other major work is that of manufacturers performed for environmental and seismic qualification of transmitters. However, the transmitter qualification data are not sufficient to address normal aging.

The aging studies reported herein used accelerated aging to accommodate the short duration of the project. Since accelerated aging does not necessarily simulate normal aging, the aging results in this report must be viewed as preliminary. Furthermore, we shall point out that this study was concerned with the performance of the portion of the pressure sensing system and electronics that are located in the harsh environments of the plant. That is, the power supply and other components of the pressure sensing channel that are located in the control room, cable spreading room, or other mild environments were not studied.

The word aging is used in this report to refer to normal operational aging which occurs with long term exposure to normal plant conditions.

2. PROJECT OBJECTIVES

The goal of this project is to provide the foundation for a study of the consequences of normal aging on the dynamic performance of safety system pressure transmitters in nuclear power plants. This is needed to ensure that the current testing methods, regulatory requirements, and industry standards and practices are adequate to track age-related degradation. The project examined the validity of the available methods for response time testing of pressure transmitters and reviewed the historical data for evidence of performance degradation problems or trends.

Current response time testing and calibration intervals for pressure transmitters are based on refueling schedules apparently for two reasons:

- A method is not available for on-line calibration of pressure transmitters, and until recently, response time testing could not be performed on-line.

- A reliable data base of degradation rates and trends is not available to justify testing on periods longer than once every refueling outage.

While testing based on refueling intervals may be adequate, there is concern that the rate of performance degradation of pressure transmitters may increase as the current generation of plants becomes older. Furthermore, on-line testing methods based on new technologies are becoming available to permit more frequent testing of transmitters and to predict incipient failures. These considerations have motivated research such as that reported herein to ensure that practical test methods and objective test schedules are used to verify proper and timely performance of safety-system pressure transmitters in nuclear power plants.

3. BACKGROUND

Pressure transmitters provide a majority of important signals that are used for control and monitoring of the safety in nuclear power plants. Depending on the plant, there are about 50 to 200 pressure transmitters in the safety systems of each plant, with the newer plants generally having the larger number of transmitters. These transmitters are tested periodically to identify and resolve any performance problems. The periodic tests are performed at each refueling outage which occurs about every 14 to 22 months depending on the plant. On-line surveillance tests and instrument channel checks which exclude the sensors are performed more frequently while the plant is operating. The tests at refueling outages include calibration of transmitters which is performed in all plants and response time testing which is performed in about 50 percent of the plants in the United States. The response time tests cover about 30 to 60 pressure transmitters, depending on the plant, and calibration includes a larger number of transmitters. A listing of pressure transmitters that are usually tested for response time is given in Table 3.1.

The interest in response time testing of pressure transmitters in nuclear power plants began when the first draft of Regulatory Guide 1.118 was issued by the NRC in the mid-1970's[1]. In response to this regulatory guide, the Electric Power Research Institute (EPRI) launched two projects to develop practical methods for response time testing of pressure transmitters. One project was performed by the Nuclear Services Corporation and another by Babcock and Wilcox Company (B&W). The work at Nuclear Services Corporation provided the equipment to perform the "substitute process variable" or ramp test[2]. The equipment produces a ramp pressure signal that results from controlled leakage of air from a high pressure cylinder to a low pressure cylinder. The equipment is referred to as the "Hydraulic Ramp Generator" and is used in many nuclear power plants to perform the required tests.

The work at B&W concentrated on the applicability of noise analysis for sensor response time testing[3]. This project and work performed by others in the late 1970's concluded that noise analysis was more suited for response time degradation monitoring than for quantitative response time testing. Recent research, however, has concluded that quantitative response time measurements can be performed using the noise analysis method.

In other related developments, a method was developed in the mid-1980's for on-line testing of response time of force-balance pressure transmitters. The method is called the power interrupt (PI) test[4]. Like noise analysis, the PI test is a passive method that can be performed on force-balance transmitters at anytime while the plant is at normal operating conditions.

TABLE 3.1

Examples of Pressure Transmitters Tested
for Response Time in Nuclear Power Plants

Pressurized Water Reactors
(PWRs)

1. Containment Pressure
2. Pressurizer Level
3. Pressurizer Pressure
4. Reactor Coolant Flow
5. Refueling Water Storage Tank (RWST) Level
6. Steam Flow
7. Steam Generator Level
8. Steam Pressure
9. Turbine Impulse Pressure

Boiling Water Reactors
(BWRs)

1. Drywell Pressure
2. Main Steam Flow
3. Reactor Vessel Pressure
4. Reactor Vessel Water Level
5. Reactor Water Clean Up (RWCU) Flow

4. TEST EXPERIENCE IN NUCLEAR POWER PLANTS

Response time testing of pressure transmitters has been performed in nuclear power plants for over ten years. The tests have resulted in response time values in the range of about 0.05 to 2.5 seconds compared to response time requirements in the range of 0.5 to 2.0 seconds. Most of the tests completed to date in nuclear power plants have used the ramp test method. Although this method can produce accurate results if performed properly, inherent difficulties in implementing the tests can render the results invalid. For example, if the test signal is oscillatory or if the transmitter is underdamped, the test output will oscillate and cause the response time results to depend on the time at which the response time is measured. This and other problems have caused some of the historical response time results to be unreliable and not useful for trending purposes. In addition, the effects of sensing line delays on the overall system response time are not addressed by the conventional methods. In fact, the sensing lines in most plants are not tested except in the process of trouble shooting calibration problems.

An informal review of limited historical results has indicated that there are not good correlations between a transmitter's response time and its manufacturer, service, or age of the plant in which it is used. In some plants, identical transmitters used for identical service have had response times that were different by as much as a factor of five. It is not known if the differences in response times of identical transmitters are due to degradation, manufacturing tolerances, calibration or test methods. The only correlation that could be found is one between response time and pressure range. It appears that high pressure transmitters have a faster response time than low pressure transmitters.

In a group of seven plants with identical transmitters, two plants were found to have average response times which were consistently larger than the other plants by a factor of about two. These two plants are older than the other plants, and it is not known if the age of the plants can account for the larger response times. This observation will have to be confirmed when more data becomes available.

5. CHARACTERISTICS OF PRESSURE TRANSMITTERS

Nuclear plant pressure transmitters are complex electro-mechanical systems designed for measurements of pressures from a few inches of water to about 3000 pounds per square inch (psi). A pressure transmitter may be viewed as a combination of two systems: a mechanical system and an electronic system. The mechanical system of a pressure transmitter contains an elastic sensing element (diaphragm, bellows, Bourdon tube) which flexes with pressure. The movement of this sensing element is detected and converted into an electrical signal proportional to the applied pressure.

Two types of pressure transmitters are used in most safety-related pressure measurements in nuclear power plants. These are referred to as motion-balance and force-balance, depending on how the movements of the sensing element are converted into an electrical signal.

In motion-balance transmitters, the displacement of the sensing element is measured with a strain gage or a capacitive detector and is converted into an electrical signal that is proportional to pressure. An example of this type of transmitter is one that consists of an oil filled cavity with a capacitance plate as the pressure sensing element. As differential pressure changes, the differential capacitance of the sensing element will change. This capacitance change is measured, amplified, and linearized by an electronic circuit into an electrical current which is the output of the transmitter. This output signal is transmitted by the same wires that provide power to operate the transmitter (i.e., the instrument is a two wire pressure transmitter). The output of this and most other nuclear-type transmitters is a DC current in the range 4 to 20 or 10 to 50 milliamperes.

In force-balance transmitters, a position-detection device is used to detect the displacement of a diaphragm, and a force motor is used to null the displacement as it develops. A feedback control system uses the displacement signal to control the force-motor operation. The force-motor current provides an electrical signal that is related to pressure.

The transmitter electronics consists of circuitry to provide signal conditioning, temperature compensation, and linearity adjustments to the output signal. The circuit has various active and passive components. The transmitter's power supply is usually located in a remote location such as the control room or cable spreading room. As such, the power supply is not usually included in performance aging studies as it is not subject to a harsh environment as are the other components of the transmitter.

The transmitter electronics for low and high pressure applications are typically the same while the sensing element is different. For example, one manufacturer uses three different elastic elements to accommodate several pressure ranges from 0 to a maximum of 3000 psi using the same transmitter housing design. Some transmitters are equipped with a damping potentiometer to reduce output noise as desired. The response times of these transmitters are thus dependent on the damping adjustment. Depending on the selected damping, typical response times of these transmitters are in the range of 0.2 to 2.0 seconds.

Four manufacturers provide most of the pressure transmitters that are used in the safety systems of nuclear power plants. These are Barton, Foxboro, Rosemount, and Tobar (formerly called Westinghouse or Veritrak). A listing of transmitter models that are tested for response time in nuclear power plants is given in Table 5.1. Note that there are a few transmitters that do not have environmental qualifications, even though they are used for safety-related applications and are response time tested. These are safety-related transmitters that are located in the areas of the plant which are not subject to the consequences of a Loss of Coolant Accident (LOCA).

The manufacturer's specifications for response time of most of these transmitters are given in Table 5.2. These response time specifications are believed to be general estimates of nominal response times of the transmitters at laboratory conditions. The in-service response time of identical transmitters may be significantly different. The response time information in Table 5.2 was obtained from general manufacturer's literature or by discussions with manufacturers' technical representatives. Note that these response time specifications have different definitions depending on the manufacturer. For example, Barton uses the time for 10% to 90% of step, and Rosemount uses time constant (time required for the sensor output to reach 63.2% of its final value after a step change in pressure). The response time specification for Foxboro transmitters is based on frequency response data. For Tobar transmitters, the response is defined by the manufacturer as the time to reach 50% of calibrated range upon application of a step change in input pressure. These different definitions result in different and often unequal indicies for expressing the reference response time of pressure transmitters.

A photograph and a simplified schematic of a typical nuclear-type pressure transmitter from each manufacturer is shown in Figures 5.1 through 5.4. (Permission to publish the photographs of these transmitters has been give to AMS by the respective transmitter manufacturers.) The Foxboro transmitters are force-balance, Rosemount transmitters are differential capacitance type, and Barton and Tobar are strain gage transmitters. The nuclear-type transmitters from the same manufacturer usually have the same physical appearance for different ranges and different levels of output currents.

TABLE 5.1

Examples of Transmitter Models
Tested for Response Time in Nuclear Power Plants

Manufacturer	Model Number	Environmental Qualification
Foxboro	• E11	No
	• E13	No
	• NE11	Yes
	• NE13	Yes
Barton	• 752	No
	• 763	Yes
	• 764	Yes
Rosemount	• 1152	Yes
	• 1153	Yes
	• 1154	Yes
Tobar	• 32 DP1	Yes
	• 32 DP2	Yes
	• 32 PA1	Yes
	• 32 PA2	Yes
Veritrak	• 76 DP1	Yes
	• 76 DP2	Yes
	• 76 PA1	Yes
	• 76 PA2	Yes

Note: Tobar Models 32 DP1, 32 DP2, 32 PA1, and 32 PA2 correspond and are
identical to Veritrak Models 76 DP2, 76 DP1, 76 PA2, and 76 PA1
respectively. Note that 32 DP1 corresponds to 76 DP2 and so on.

TABLE 5.2

Manufacturer's Specifications for Response
Time of Some Nuclear Plant Pressure Transmitters

Manufacturer	Model	Range	Response Time (sec)
Barton	763	All	<0.18
	764	All	<0.18
Foxboro	E11	All	<0.30
	E13	All	<0.30
	NE11	All	<0.30
	NE13	All	<0.30
Rosemount	1152	3	0.31
		4	0.13
		5	0.09
		6	0.06
	1153	3	2.0
		4	0.5
		5 - 9	0.2
	1154	4	0.5
		All Others	0.2
Tobar	32 DP1	0-600"	0.08-0.15
Veritrak	76 DP2	0-600"	0.08-0.15

Notes: 1. *Above results are general estimates of nominal response times of the transmitters at laboratory conditions. The in-service response times of identical transmitters may be significantly different than the values given in this table.*

2. *The above response time data from different manufacturers are not based on the same definition. Therefore, these values should not be used for comparison of sensors from different manufacturers.*

Photograph of a Barton Transmitter

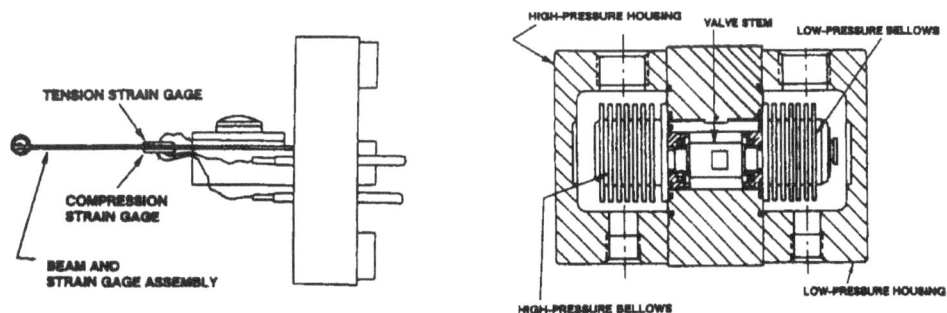

Figure 5.1.   Photograph and Schematic Diagram of a Barton
Pressure Transmitter.

Photograph of a Foxboro Transmitter

AMS—DWG PXT029C

Vector Flexure

Span Adjustment

Detector Armature (Ferrite Disk)

Differential Transformer

Force bar

Amplifier and Span Selector

Force Feedback Motor

Receiver

Power Supply

Feedback Force Lever

Zero Adjustment

Diaphragm Seal

Diaphragm Capsule

Connecting Link

Figure 5.2. Photograph and Schematic Diagram of a Foxboro Pressure Transmitter.

Photograph of a Rosemount Transmitter

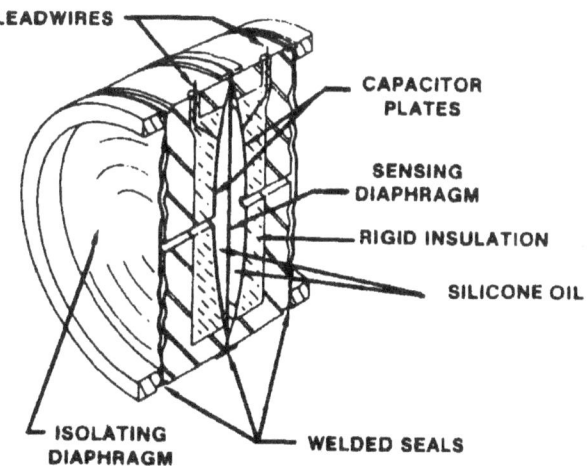

LEADWIRES

CAPACITOR PLATES

SENSING DIAPHRAGM

RIGID INSULATION

SILICONE OIL

ISOLATING DIAPHRAGM

WELDED SEALS

Figure 5.3. Photograph and Schematic Diagram of a Rosemount Pressure Transmitter.

Photograph of a Tobar Transmitter

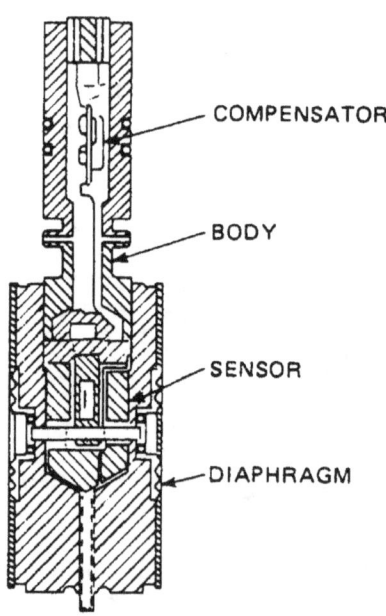

COMPENSATOR

BODY

SENSOR

DIAPHRAGM

Figure 5.4. Photograph and Schematic Diagram of a Tobar
Pressure Transmitter.

6. CHARACTERISTICS OF SENSING LINES

Sensing lines, also referred to as impulse lines or instrument lines, are used to connect the process medium to the transmitter (Figure 6.1). At normal operation, there is no flow through the sensing lines. The transmitter is located away from the process to minimize temperature, vibration, and other effects and to facilitate physical access to the transmitter. Temperature effects are especially important and must be kept at a suitable level to preserve the qualification of the transmitters for nuclear service.

Temperature effects are also important in such measurements as containment pressure where appropriate steps must be taken to prevent temperature variations in the reference leg of the transmitter from affecting the output. For this reason, in some plants, containment pressure transmitters are located outside the containment to guard against the effects of temperature on pressure measurements.

Sensing lines are usually made of 1/2-to 3/8-inch stainless steel tubing. They are designed to allow for thermal expansion and vibration without deformation, to ensure drainage by gravity, and to provide for self venting. For fluid filled sensing lines which are the subject of interest in this study, self venting is accomplished by sloping the sensing line downward to allow any gas or air in the line to vent to the process. The slope for sensing lines is typically about one inch per foot. When this is not possible, the sensing line is sloped as much as possible but usually not less than about 1/8 inch per foot.

Depending on the plant's physical layout, sensing lines are about 20 to 200 feet long with most less than 100 feet long. The length of the sensing line is usually kept to a minimum for optimum response time. Sensing lines that are free of obstructions or voids do not add a noticeable delay to the overall system response time. However, numerous cases of blockages and voids in sensing lines that can cause significant transient delays have been reported in nuclear power plants.

For non-safety-related applications, multiple transmitters sometimes share a common sensing line, but for safety system measurements, only one transmitter is usually installed on a sensing line to avoid common mode problems such as sensing line blockages, root valve failures, etc. Another practice on non-safety-related sensing lines which is not used on safety-related sensing lines is the use of snubbers or pulsation dampers to reduce process noise to obtain good control or indication. A disadvantage of these dampers is that they increase the effective response time of the pressure sensing system.

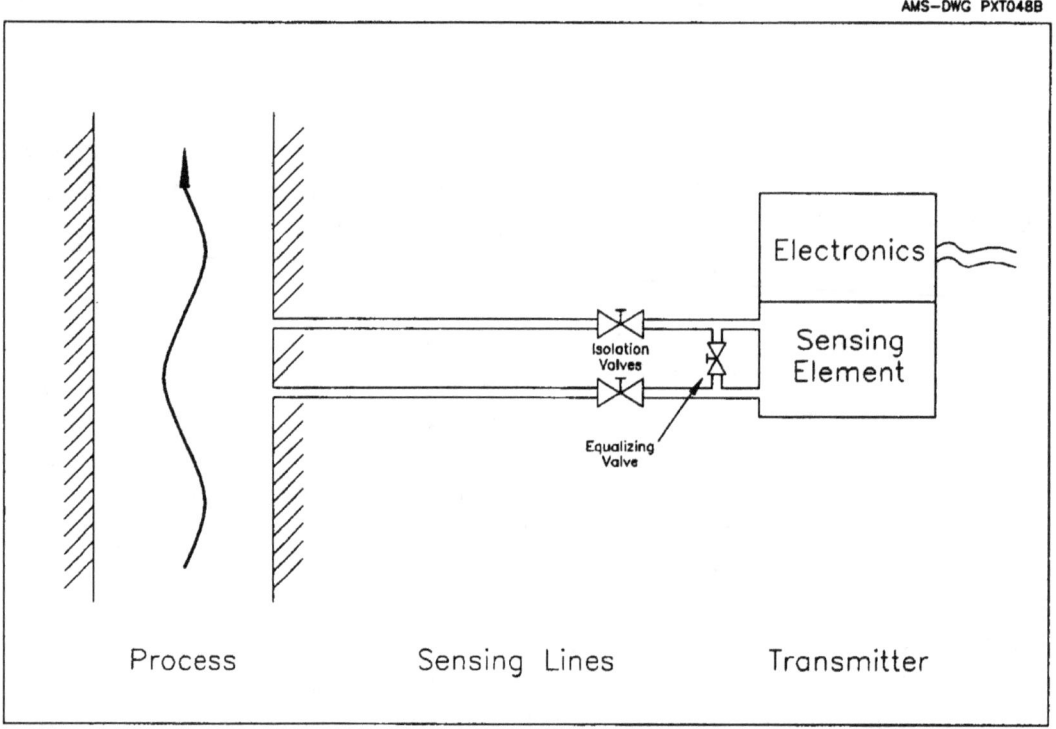

Figure 6.1. Components of a Typical Nuclear
Plant Pressure Sensing System.

7. CAUSES AND MECHANISMS OF DEGRADATION

Pressure sensing systems in nuclear power plants are exposed to conditions that can cause performance degradation over a long period of exposure. The conditions include:

- Environmental conditions such as temperature, humidity, vibration, radiation, and fluctuations in transmitter power supply voltages.

- Stresses due to changes in normal operating conditions and plant trips that result in temperature and pressure cycling.

- Degradation induced during calibration, testing or maintenance such as over-pressurization, injection of test signals to the wrong side of the transmitter, etc.

Table 7.1 provides a listing of general stress factors that can cause degradation in pressure transmitters. This is followed by Table 7.2 with a listing of failure mechanisms[5]. Descriptions of the stress factors and their effects on performance of pressure sensing systems are discussed in the following sections.

7.1 STRESS FACTORS AND THEIR EFFECTS

At normal operating conditions, pressure sensing systems are exposed to a variety of conditions that can cause performance degradation over a long period of time. Some of these factors are described below:

- Temperature. Temperature is the dominant stressor and it predominantly affects the transmitter's electronics. The ambient temperature in the reactor containment is about 120°F ± 20°F during normal operating conditions. Long term exposure to such temperatures is detrimental to the life of the transmitter. Temperature also affects other stressors. For example, detrimental effects of humidity are increased at higher temperatures because of higher diffusion rates at elevated temperatures. Figure 7.1 presents an example of the qualified life of a transmitter as a function of ambient temperature.

- Pressure. Pressure transmitters are continuously exposed to small pressure fluctuations during normal operation and large pressure surges during reactor trips and other events. Water hammer, for example, is a well-known phenomenon in nuclear power plants which can degrade the performance of pressure transmitters. Other pressure induced degradations may occur

TABLE 7.1

Examples of Potential Stress Factors

1. Temperature

   - high ambient temperatures
   - ambient temperature transients and cycling
   - temperature changes inside the transmitter due to self heating

2. Pressure

   - high process pressure
   - process pressure cycling

3. Humidity

   - high and low ambient humidity
   - high and low internal humidity

4. Vibration

   - mechanical vibration during normal operation
   - vibration during plant trips

5. Maintenance

   - repair and maintenance of circuit board
   - calibration and response time testing
   - vent/drain valve cycling

6. Transmitter Power Supply

   - voltage fluctuations
   - high output voltage

7. Other

   - radiation
   - chemical composition of ambient atmosphere
   - dirt and deposits in sensing lines

TABLE 7.2

Examples of Failure Mechanisms in Pressure Sensing Systems

1. Thermal expansion

2. Oxidation

3. Metal strain, corrosion, and fatigue

4. Plastic deformation

5. Radiation energy absorption

6. Polymerization and depolymerization

7. Outgassing

8. Failure of semi-conductors

9. Wear

10. Obstruction or clogging of sensing lines

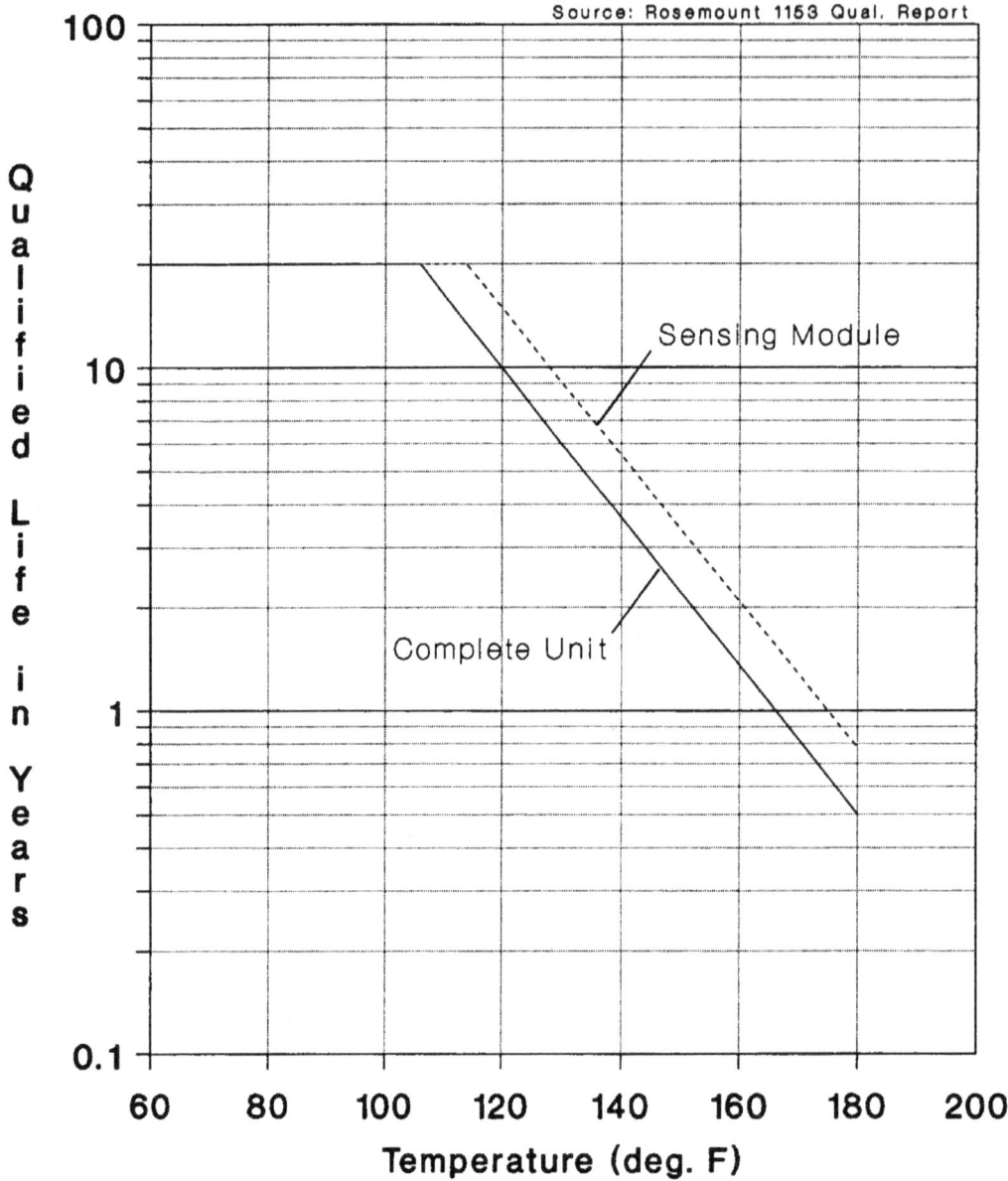

Figure 7.1.   Illustration of Qualified Life of A Typical Pressure
Transmitter as a Function of Ambient Temperature.

during calibration and maintenance when transmitters are inadvertently pressurized or cycled with pressures that are above or below their normal range.

- Humidity. Humidity affects the operation of a transmitter's electronics. Moisture sources and sinks exist within the transmitter and are therefore unavoidable. The humidity levels inside reactor containment are in the range of 10 to 100 percent. Some humidity will leak into the sensor because the organic polymer seals cannot provide hermeticity under long-term exposure to the temperatures that exist around pressure transmitters[6]. A significant degrading effect of humidity is short circuits in the transmitter electronics.

- Vibration. Vibration generated by nearby machinery during plant operation is transmitted to pressure sensors through the building structure. The vibration of concern in this aging project is not that of seismic events addressed during the qualification of pressure transmitters.

- Maintenance. Maintenance is one of the important causes of degradation in many components of nuclear power plants. In the case of pressure transmitters, calibration is performed on almost all pressure transmitters during refueling outages. Since calibration often drifts between outages, the "zero" and "span" potentiometers are adjusted frequently. This will cause the components to eventually wear out. Another example of a maintenance induced problem occurs when test pressures are applied to the wrong side of the transmitter or when isolation and equalizing valves are not manipulated in the correct sequence to prevent exposure of the transmitter to sudden changes in pressure.

7.2 EFFECT OF STRESS ON DYNAMIC PERFORMANCE

The stress experienced by a pressure transmitter during normal plant operation can cause performance degradation in the mechanical system and/or the transmitter electronics as discussed below. Some examples are:

- Permanent deformation of sensing elements or the mechanical linkages due to pressure surges during reactor trips and maintenance.

- Failure of the bellows. Bellows can rupture and cause leaks, false pressure indications, and sluggish response.

- Degradation or leakage of fill fluid. The fill fluid (usually oil) in pressure transmitters can suffer degradation or leak out. If the degradation involves changes in fluid properties, changes in response time may result. Any leakage of the fill fluid may be accompanied by degradation of response time and calibration. Recent incidents of loss of fill fluid in safety system pressure transmitters have been encountered in several nuclear power plants.

• Failure of diaphragm due to work hardening. Work hardening may cause cracks or fatigue in the diaphragm and change its stiffness.

• Friction in mechanical linkages causes response time degradation (it may or may not have any effect on calibration).

• Failure of seals. Seals can harden, crack, or take a set, allowing moisture to leak into the transmitters.

• Loosening of mechanical components in force balance transmitters due to pressure fluctuations, surges, and mechanical vibration.

The electronic components of pressure transmitters include numerous resistors, capacitors, diodes, and integrated circuits that are used for signal conversion, signal conditioning, and linearization of the transmitter's output. In some transmitters, 10 to 20 resistors are used to maintain the linearity of the sensor output in addition to resistors and capacitors to set the transmitter "zero" and "span". Almost all these components are strongly affected by long term exposure to temperature and humidity. To a lesser degree, they are affected by radiation and fluctuations or step changes in the power supply voltage. Any change in the value of electronic components such as the resistors or capacitors can cause calibration shifts and response time changes and also affect the linearity of the sensor output signal.

A summary of potential degradation effects on pressure transmitters that can cause response time problems is given in Table 7.3.

7.3 PERFORMANCE DEGRADATION OF SENSING LINES

Sensing lines are not as susceptible to aging degradation as pressure transmitters. However, there are situations that can lead to increased response times resulting from problems in sensing lines. Examples of sensing line problems that can result in sluggish response times of a pressure sensing system are:

• Blockages due to sludge, boron, or deposits
• Air or gas entrapped in the sensing line
• Freezing of sensing lines due to problems with heat trace on the lines
• Improper line-up or seating of isolation and equalizing valves
• Leakage in sensing lines due to valve problems

TABLE 7.3

Examples of Effects That Can Cause
Response Time Degradation in Pressure Transmitters

Degradation	Cause	Effect
Deformation of diaphragm	Pressure fluctuations pressure surges and mechanical vibration	Changes in stiffness of sensing element
Wear and friction of mechanical linkages	Pressure fluctuations and surges, corrosion and oxidation	Changes in system restoration ability
Partial or total loss of fill fluid	Manufacturing flaws Improper handling	Significant capacitance changes Nonlinear output
Degradation of fill fluid	Chemical changes of oil due to radiation and/or heat	Viscosity Changes
Failure of seals	Embrittlement and cracking	Moisture on electronics
Leakage of process fluid into cell fluid	Cracking of diaphragm	Capacitance changes
Changes in values of electronic component	Heat, radiation, humidity, voltage stresses, and maintenance	Changes in dynamic response and linearity of electronics
Hysteresis	Pressure fluctuations and surges and mechanical vibration	Distorted output
Setpoint drift	Calibration shifts	Increased time to reach setpoint

Any combination of the above problems can increase the response time of a pressure sensing system. A number of LERs have revealed significant problems in nuclear power plants due to sluggish transmitter responses caused by sensing line blockages, voids, and similar phenomena. The presence of air in the sensing line can cause not only increases in response time but also can cause resonances which produce pressure variations and false pressure indications. Although at high pressures, air may dissolve in the fluid, there are a number of cases where entrapped air remains undissolved. In addition to causing transient response problems, air in the sensing line can affect the accuracy of pressure indications.

8. RESPONSE TIME FUNDAMENTALS

The response time of a pressure transmitter is tested using a test signal such as a step, a ramp, or a sinusoidal function. The test signal is usually selected based on what may happen in the process. In nuclear power plants, ramp signals are more often used for pressure transmitters than step signals. The response to step or ramp input signals is called step response and ramp response respectively. Associated with step response is a unique index called time constant, and with ramp response an index called ramp time delay. Time constant is defined as the time required for the sensor output to reach 63.2 percent of its final value following a sudden change in applied pressure (Figure 8.1). The ramp response, on the other hand, is the asymptotic delay between the applied pressure and the indication of the sensor (Figure 8.2).

The terms time constant and ramp time delay are two different indices for quantifying the response time of a pressure transmitter. If the transmitter can be approximated as a first order dynamic system, the ramp time delay and time constant would be numerically the same. This relationship is shown below starting with the transfer function, $G(s)$, for a first order system:

$$G(s) = \frac{1}{\tau s + 1} \tag{8.1}$$

where τ is the transmitter's time constant and s is the Laplace transform variable. The response $y(t)$ for a step input can be obtained from Eq. 8.1:

$$y(t) = 1 - e^{-t/\tau} \tag{8.2}$$

Note that for time $(t) = \tau$, Eq. 8.2 will have the following value which defines the time constant:

$$y(t) = 0.632 \tag{8.3}$$

Using the transfer function of Eq. 8.1, the response $R(t)$ to a ramp pressure input with ramp rate k may be written as:

$$R(t) = k(t - \tau + \tau e^{-t/\tau}) \tag{8.4}$$

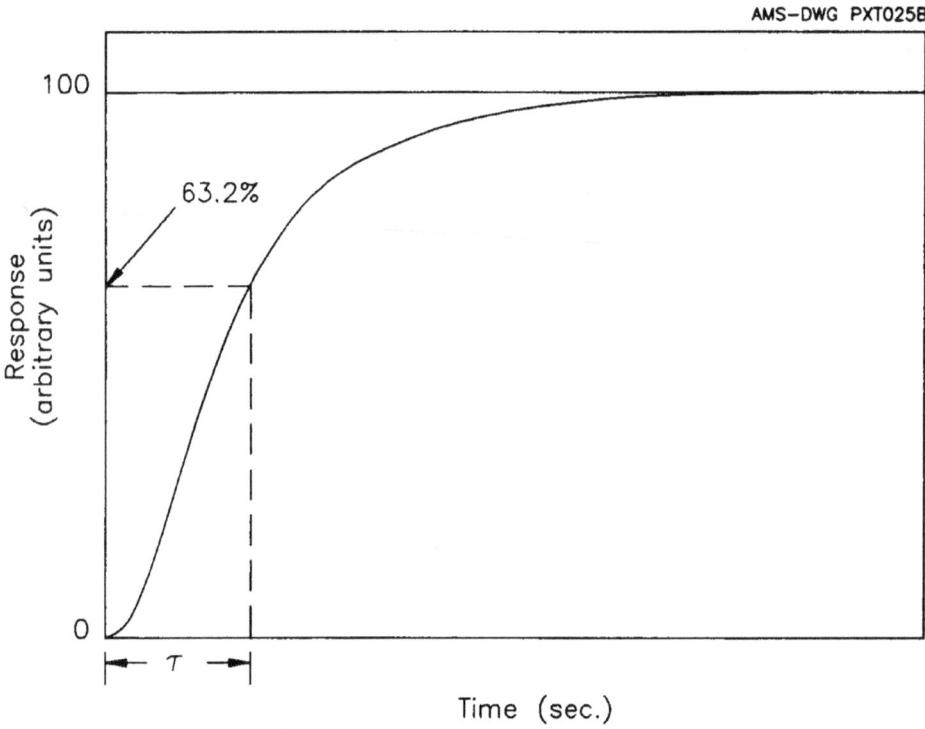

Figure 8.1.   Illustration of the Step Response and
              Calculation of the Time Constant (τ).

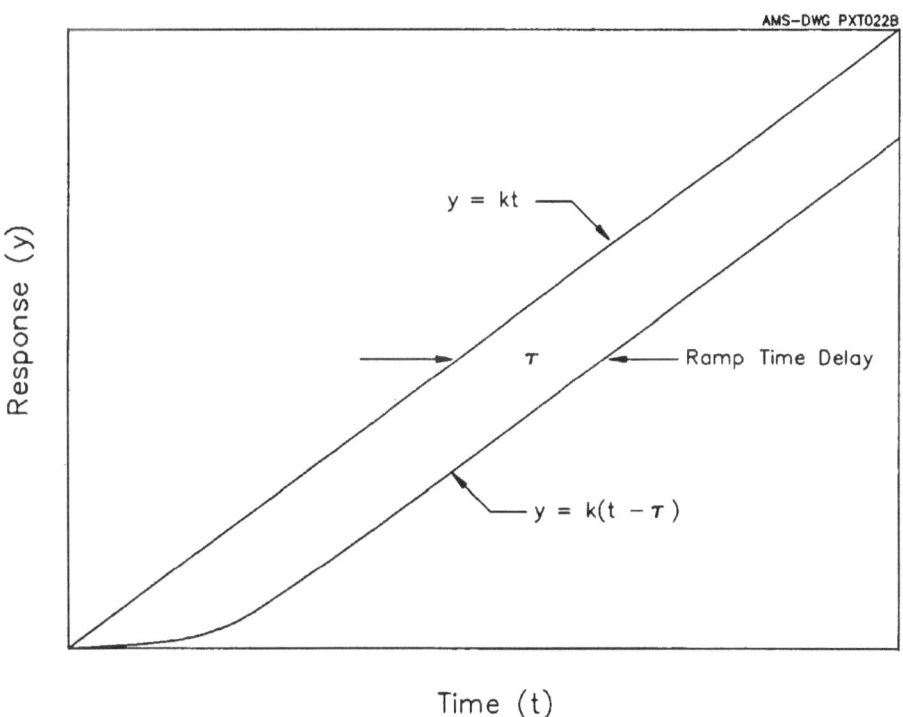

Figure 8.2. Ideal Ramp Response of a Pressure Transmitter.

At $t >> \tau$ we can write:

$$R(t) = k(t - \tau) \tag{8.5}$$

That is, the ramp time delay is equal to τ when enough time is elapsed for the exponential portion of response to decay (Figure 8.2).

In the frequency domain, it can be shown that the reciprocal of the break frequency (F_b) in the frequency response plot of a pressure sensor is also equal to τ for a first order system (Figure 8.3). Starting with Eq. 8.1 and substituting $j\omega$ for s, we can write

$$G\ (j\omega) = \frac{1}{j\omega + p} \tag{8.6}$$

where $p = 1/\tau$ is referred to as the system pole and:

$$j = \sqrt{-1}$$
$$\omega = \text{angular velocity in radians per second.}$$

The gain of this transfer function is then:

$$|G| = \begin{cases} 1/p & @\ \omega = 0 \\ (0.707)\ (1/p) & @\ \omega = p \end{cases} \tag{8.7}$$

Therefore:

$$|G|_{\omega = p} = 0.707\ |G|_{\omega = 0} \tag{8.8}$$

Eq. 8.8 shows that the pole "p" can be found from the gain plot (Figure 8.3) at the frequency where the gain is 0.707 below the low frequency gain.

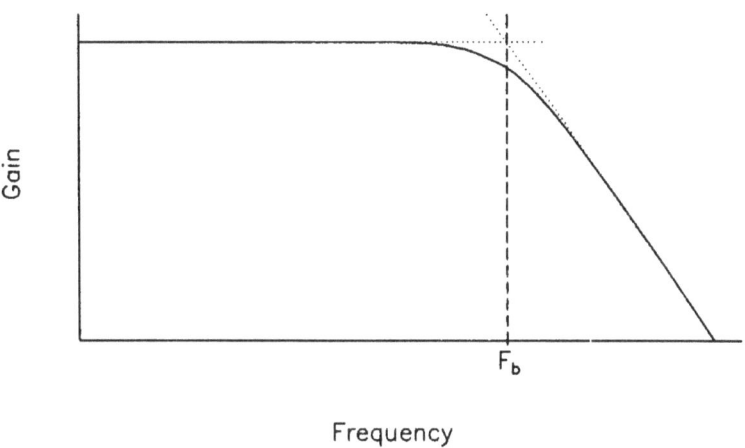

Figure 8.3.   Frequency Response of a First Order System.

We must point out, however, that pressure sensors are not generally first order. Nevertheless, the term time constant as defined above is used to describe pressure sensor response time regardless of dynamic order. Moreover, the response time of pressure sensors in some plants is determined in terms of a time to reach a setpoint following a step pressure input to the sensor. This is referred to as "time-to-trip". Based on the results of laboratory experiments with typical pressure transmitters and certain assumptions, the values for time-to-trip and for time constant can generally be viewed as conservative estimates of ramp time delay except for cases of underdamped systems.

If a pressure transmitter's dynamics are to be treated properly, a general n^{th} order response such as the one that follows should be used to represent the output of the transmitter to a step input:

$$y(t) = A_0 + A_1 e^{-t/\tau_1} + A_2 e^{-t/\tau_2} + \ldots + A_n e^{-t/\tau_n} \qquad (8.9)$$

where n is the order of the system and,

$$A_0, A_1, A_2, \ldots, A_n = constants$$
$$\tau_1, \tau_2, \tau_3, \ldots, \tau_n = modal\ time\ constants\ (the\ time\ constant$$
$$for\ the\ i^{th}\ term\ or\ mode\ in\ the\ solution).$$

Under certain assumptions, the overall time constant (τ) of the system may be approximated in terms of the modal time constants ($\tau_i's$) as,

$$\tau = \tau_1 \left[1 - ln \left(1 - \frac{\tau_2}{\tau_1} \right) - ln \left(1 - \frac{\tau_3}{\tau_1} \right) - \ldots \right] \qquad (8.10)$$

where "ln" represents the natural logarithm. The ramp time delay (τ_d) of the system of Eq. 8.9 may be then written as:

$$\tau_d = \tau_1 + \tau_2 + \ldots + \tau_n \qquad (8.11)$$

Note that both the time constant (τ) and ramp time delay (τ_d) can be obtained from the modal time constants for the system considered here. Thus, if we have the step response (Eq. 8.9) of a sensor, we can identify the modal time constants (τ_i) and use them in Eq. 8.11 to calculate the ramp time delay.

9. RELATION OF RESPONSE TIME WITH CALIBRATION

The effective response time of a pressure transmitter may be viewed as the sum of two separate components: an intrinsic response time (τ_I) and a setpoint delay (τ_D). These are illustrated in Figure 9.1. The transient on the top of Figure 9.1 illustrates the effective response time of a transmitter that is in perfect calibration, and the transient on the bottom illustrates the effective response time when a positive "zero" or gain shift has occurred in the transmitter's calibration. Note that the amount of delay caused by the setpoint drift depends on the ramp rate. It is apparent that if the setpoint shift is negative, the effective response time will be less than the intrinsic response time.

The intrinsic response time is the parameter that we measure during a response time test of the pressure transmitter. This parameter depends on electro-mechanical operation of the transmitter and can change with changes in mechanical or electrical characteristics of transmitter's components. These changes may or may not affect the calibration of the transmitter. However, a clear relationship between the intrinsic response time of a transmitter and its calibration cannot be readily envisioned. Experience has shown that only gross malfunctions may manifest themselves in both calibration and response time and that changes such as a few seconds in response time may not be accompanied by corresponding calibration shifts and vice versa.

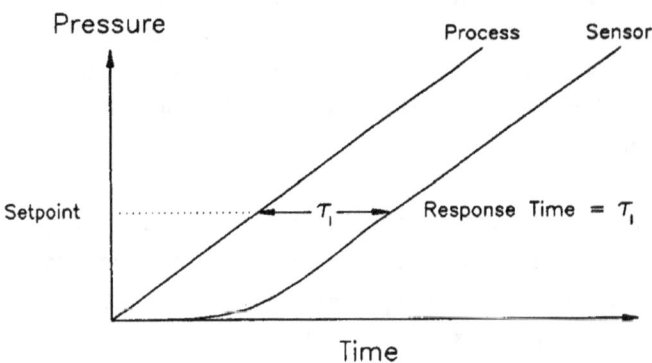

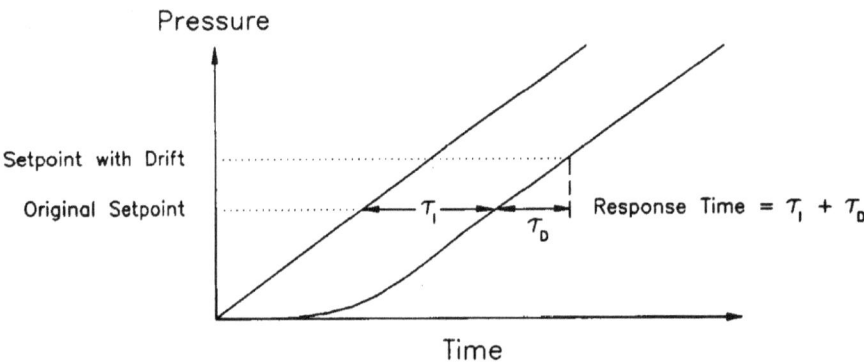

Figure 9.1. Illustration of Ramp Response of a Pressure Transmitter With and Without a Calibration Shift.

10. RESPONSE TIME TESTING METHODS

The response time testing methods for pressure transmitters may be divided into two groups of tests: on-line methods which are based on new technologies developed in the last five years and the conventional methods which have been used since testing began about ten years ago. There are three tests (ramp, step, and frequency) which we refer to as conventional methods and two tests (power interrupt test and noise analysis) which are referred to as new methods or on-line methods. All five methods were evaluated in this project for general effectiveness in providing reliable dynamic performance information. A description of the conventional methods and the on-line methods follows.

10.1 CONVENTIONAL METHODS

The conventional tests depend on a pressure test signal which is applied to the transmitter under test, and its delay is measured with respect to a fast-response reference transmitter. While the conventional tests do not require removal of the transmitters from the plant, local access to the transmitter is required for testing. Therefore, these tests can be performed only during cold shutdown. Besides the obvious disadvantage of radiation exposure to test personnel and impact on shutdown schedules, the disadvantage of conventional methods is that the effects of operating conditions such as static pressure and temperature are not included in the test results. Furthermore, in conventional tests, the sensing lines are valved off and their effects are therefore not taken into account.

Depending on the choice of the pressure test signal, three methods are available as described below:

- Ramp Test. This test involves applying a pressure ramp signal to the transmitter under test and simultaneously to a high-speed reference transmitter (Figure 10.1). The delay between the output of the two transmitters when they reach a predetermined setpoint is measured as the response time of the transmitter under test. The method is called "substitute process variable" or ramp test. The equipment used for this test is called the "Hydraulic Ramp Generator" which was developed in the late 1970's by the Nuclear Services Corporation under a contract with the Electric Power Research Institute[2]. A simplified schematic of the equipment is shown in Figure 10.2. Most of the current tests in nuclear power plants are performed using this or similar equipment.

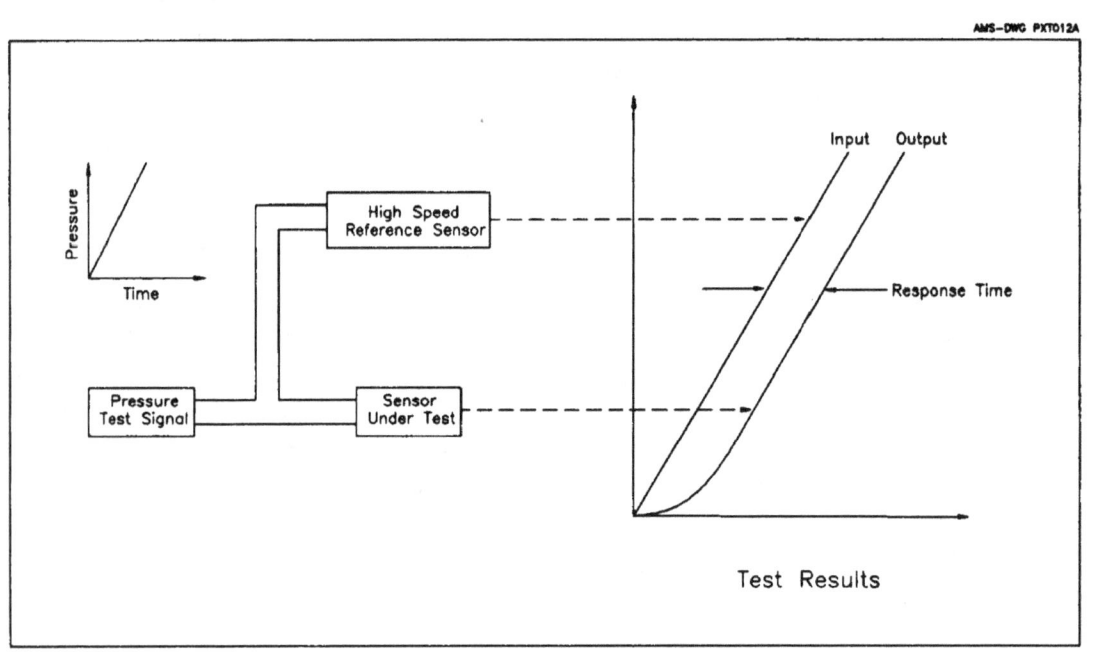

Figure 10.1. Illustration of the Ramp Test Method.

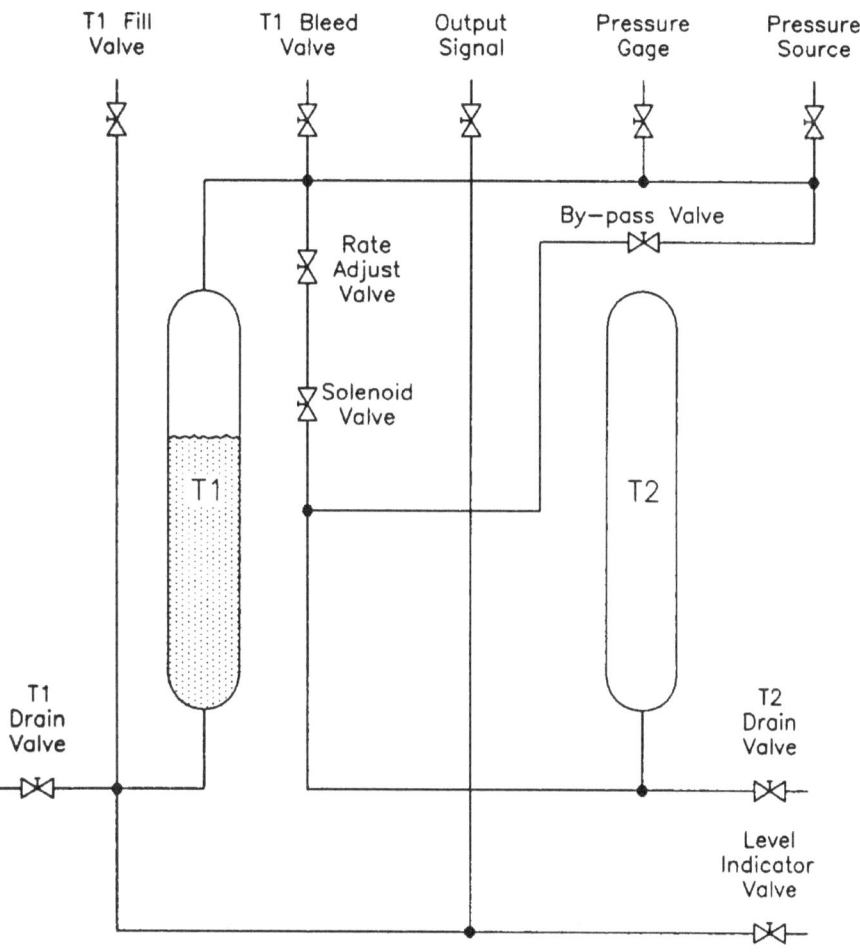

Figure 10.2. Simplified Schematic Diagram of the Ramp Test Equipment.

A ramp signal is usually used for describing the response time of pressure transmitters because design basis accidents in nuclear power plants assume pressure transients that approximate a ramp. The ramp test is the most widely used and accepted method for response time testing of pressure transmitters in nuclear power plants. Therefore, we used the ramp method as the reference method for assessment of the other conventional methods as well as the new methods. Any method that can provide the same results as the ramp method is therefore considered an acceptable alternative to the ramp test.

- Step Test. This test is similar to the ramp test except that it involves using a step pressure signal rather than a ramp pressure signal. It can be performed with the same instrument used for ramp tests or with a simpler instrument involving a pressure source and a fast-acting solenoid valve. The response time obtained from a step test is equal to the time required for the sensor output to reach 63.2 percent of its final steady state value after a step change in input. This is usually a conservative estimate of the asymptotic ramp time delay. The step test is used in a few nuclear power plants using equipment that is setup by utility personnel. The interpretation of step test results performed by utilities is usually different than those of laboratory tests where the response time is described in terms of the time to reach 63.2 percent of steady state output. The utility tests measure a quantity called "time-to-trip" which is equal to the time difference between the initiation of a step input and the time when output reaches a pre-determined setpoint at the end of the instrument channel. The advantage of this test is that it accounts for all components of the channel in a single test.

- Frequency Test. This test employs a pressure waveform generator to provide a sinusoidal shaped pressure signal. The signal is applied to a reference transmitter as well as to the transmitter under test. The outputs of the two transmitters are then used to generate a Bode plot (ratio of output to input versus frequency) from which the response time of the sensor can be estimated (Figure 10.3). The frequency test involves two different types of equipment depending on the operating range of the sensor under test. The low pressure test equipment is shown in Figure 10.4 and the high pressure equipment in Figure 10.5. For low pressure testing, this equipment provides a time varying periodic signal similar to a sinewave by driving a piston in and out of a cylinder that moves above a fluid stream. The test instrument is equipped with a transmission system to permit changing of the signal frequency. The high pressure instrument uses a current to pressure (I/P) converter to generate a time varying test signal that is amplified with a pressure amplifier. The frequency test equipment and procedures are used mostly in laboratory research and not in nuclear power plants. The high pressure frequency test equipment used here is limited in frequency response and is mainly useful for such applications as linearity testing of transmitters where a high pressure waveform may be used.

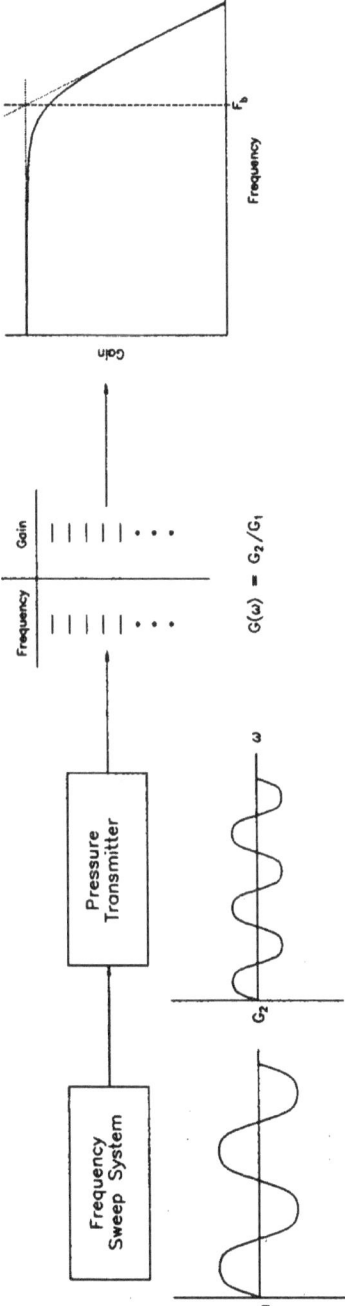

Figure 10.3. Illustration of Frequency Test Principle.

Figure 10.4. Photograph of Frequency Test Equipment
for Low Pressure Testing.

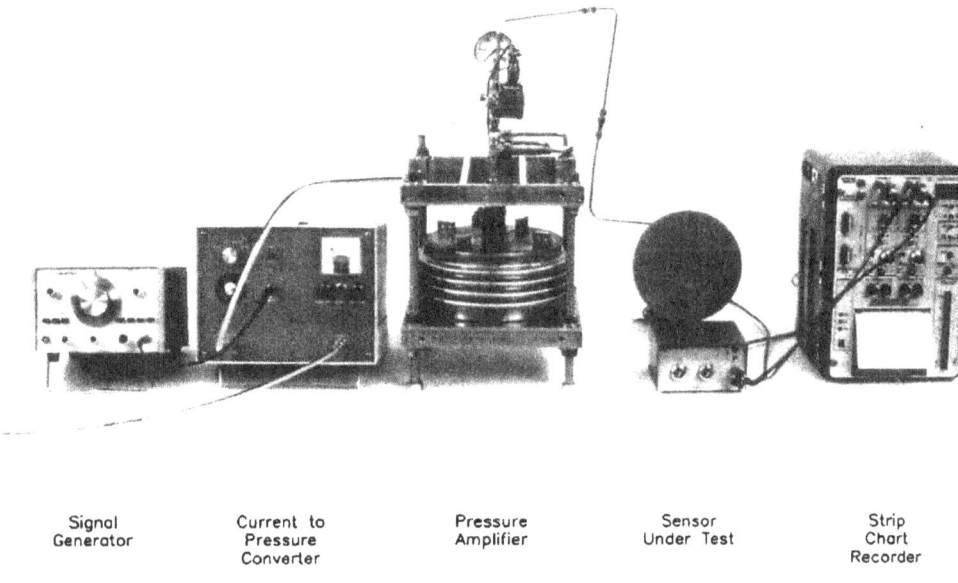

Signal	Current to	Pressure	Sensor	Strip
Generator	Pressure	Amplifier	Under Test	Chart
	Converter			Recorder

Photograph of the High—Pressure Test Setup

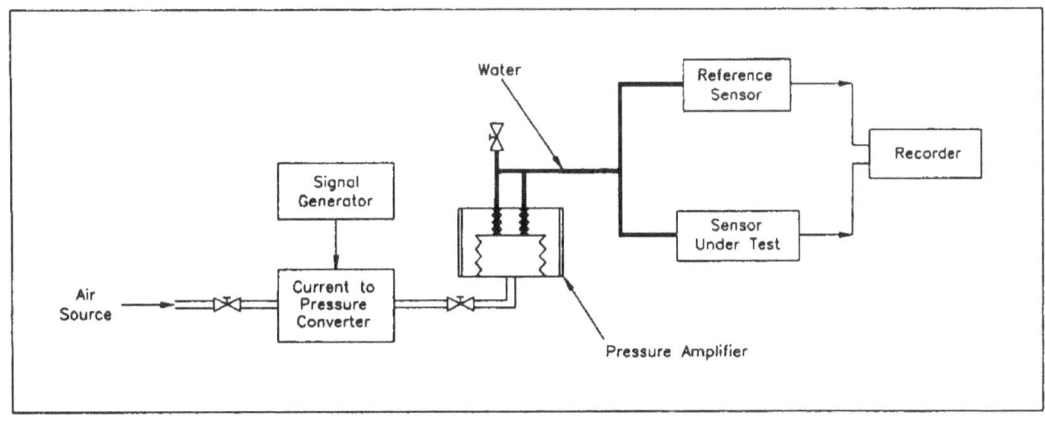

Schematic Diagram of the High—Pressure Test Setup

Figure 10.5. Photograph and Schematic Diagram of Frequency
Test Equipment for High Pressure Testing.

10.2   ON-LINE TESTS

Two methods are available for remote testing of installed pressure sensors while the plant is operating. They are referred to as the "noise analysis" method and the power interrupt or PI test.   These methods are new in that they were recently validated for quantitative measurements in nuclear power plants. These methods account for any effect of operating conditions such as the static or working pressure, temperature, etc. Therefore they measure the actual in-service response time of the transmitters. The power interrupt test is applicable to only one class of pressure transmitters while noise analysis can be used for response time testing of any pressure transmitter.

- **Noise Analysis Method.**   Noise analysis is based on monitoring the natural fluctuations that exist at the output of pressure transmitters while the process is operating. These fluctuations (noise) are usually due to turbulence induced by the flow of water in the system, random heat transfer in the core, or other naturally occurring phenomena. The noise is extracted from the sensor output by removing the DC component of the signal and amplifying the remaining components (Figure 10.6). This signal can be analyzed to provide the response time of the pressure sensing system.

  Figure 10.7 shows a sensor which exhibits a time varying output, δo, for a time varying input δI. The sensor is represented by its transfer function (G). These are related to one another as:

$$G = \frac{\delta o}{\delta I} \quad or \quad \delta o = G \, \delta I \tag{10.1}$$

  There are three components involved here: the input, the output, and the transfer function of the sensor whose dynamic characteristics are to be determined. If any two of these three components are known, the third one can be identified. In noise analysis, we can measure the output and make an assumption about the input.   The input is a random variable and cannot therefore be characterized deterministically.   Thus, we will characterize it statistically.   Eq. (10.1) may be written in terms of the power spectral density (PSD) of the input and output signals:

$$(PSD)_o = |G|^2 (PSD)_I \tag{10.2}$$

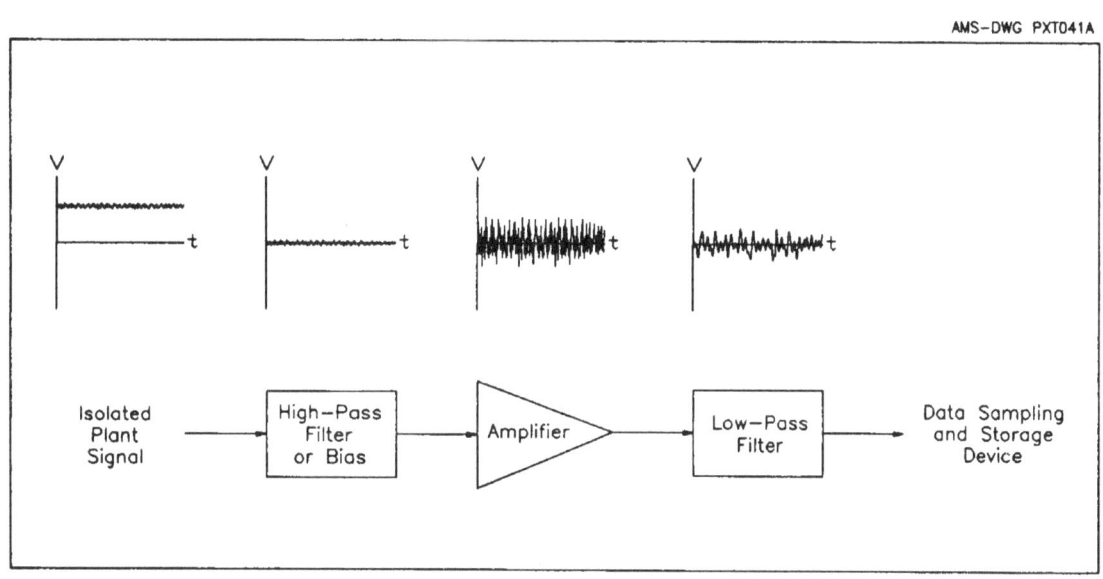

Figure 10.6.   Illustration of Noise Data Acquisition System.

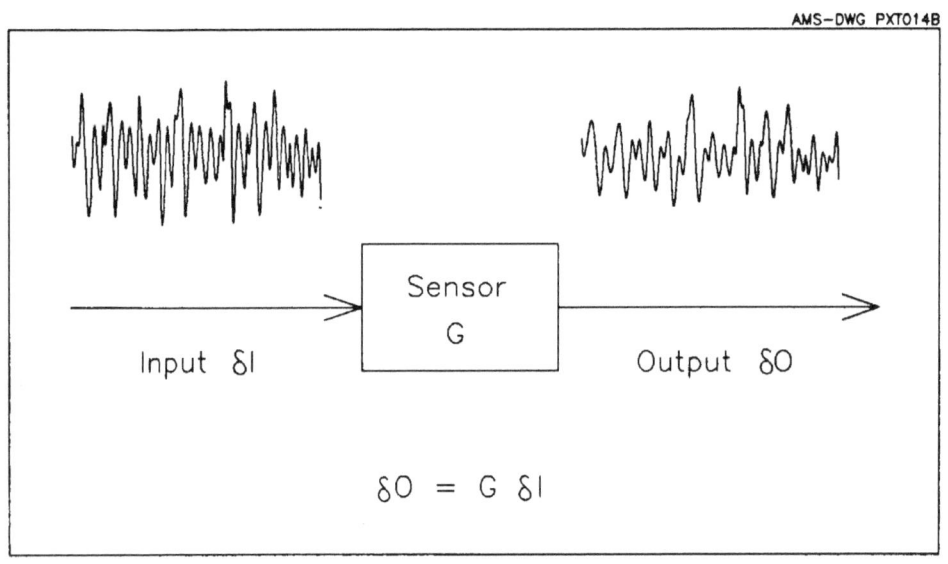

Input δI Sensor G Output δO

$$\delta O = G \, \delta I$$

Figure 10.7. Illustration of Principle of Noise Analysis.

If the process pressure is stationary and random, it would be called a white noise signal whose PSD is constant. In this case, the PSD of the output will be proportional to sensor transfer function.

The above discussions show that the PSD of the sensor output fluctuations contains the dynamic information needed for determining the sensor response time. The procedure is illustrated in Figure 10.8 for a frequency domain analysis. This analysis involves performing a Fast Fourier Transform (FFT) on the sensor output signal to obtain its PSD. A function is then fit to the PSD and the parameters of the function are identified and used to calculate the sensor response time. Noise analysis has been recently validated for quantitative response time testing of pressure transmitters in nuclear power plants.

An important advantage of noise analysis is that its results include the effect of sensing line delays and thereby gives the overall response time of the pressure sensing system.

- Power Interrupt Test. The power interrupt (PI) test is applicable to one class of pressure transmitters, that is, the force balance pressure transmitters. The PI test is based on a momentary interruption of the normal electric supply to the transmitter. The test is performed by turning the power to the transmitter OFF for a few seconds, and then ON. When the power is turned ON, the transmitter provides an output that can be analyzed to give the transmitter's response time. The test setup is illustrated in Figure 10.9. The method has been validated for testing of Foxboro pressure transmitters. Foxboro transmitters are the only force-balance type pressure transmitters that are used for safety-related applications in nuclear power plants and are subject to response time testing.

  A thorough evaluation of the operation of a Foxboro force-balance pressure transmitter has shown that the essential dynamics of the power-up operation in a PI test duplicate the transmitter's response to an external pressure step. Thus, an appropriate analysis of this response provides the essential dynamic information for determining the response time of the transmitter.

Both noise analysis and PI tests require computer analysis. The analysis yields the dynamic parameters that are used to obtain the step response, ramp response, or any other dynamic parameter of interest. Both noise analysis and PI tests have been successfully implemented in a number of nuclear power plants. These tests do not interfere with plant operations. The noise analysis have to be performed when the plant is at or near normal operating conditions, but the PI test can be performed remotely at anytime as long as the transmitters are under pressure.

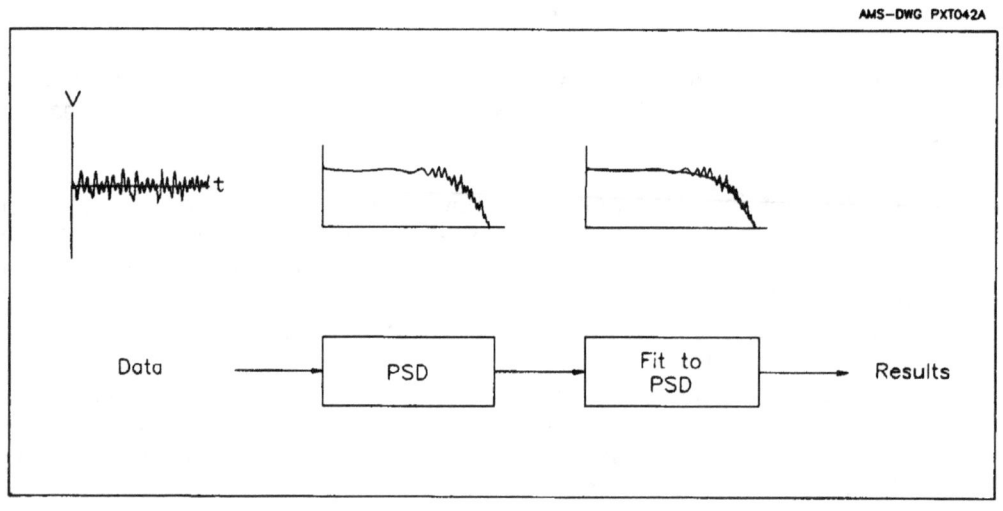

Figure 10.8.   Noise Data Analysis Procedure.

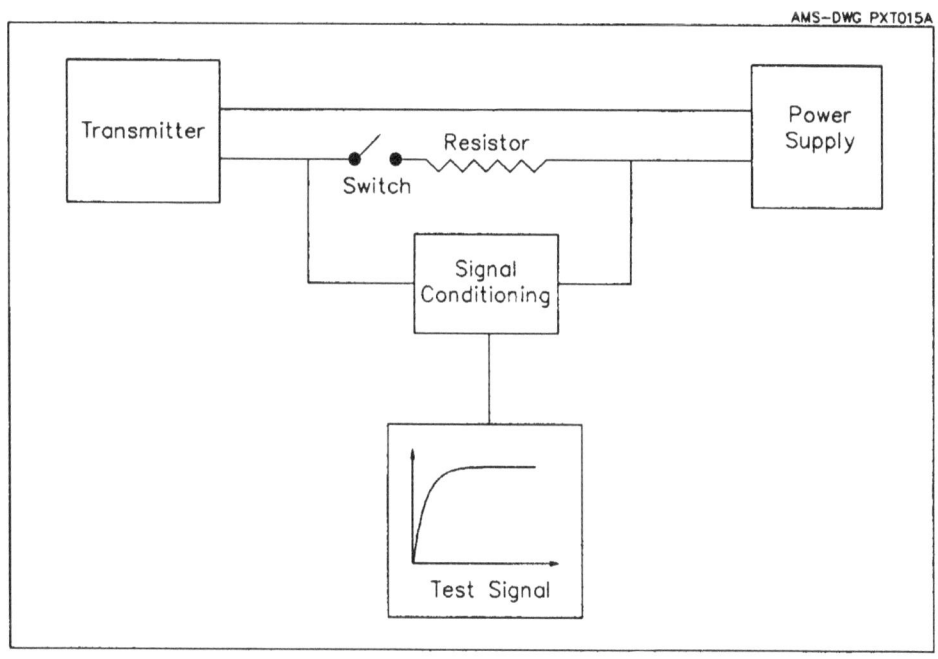

Figure 10.9.   Power Interrupt Test Setup.

11. ACCELERATED TESTING

It is well established that the conventional accelerated aging based on Arrhenius and other theories does not simulate real time normal aging of a complex system such as a pressure transmitter. However, accelerated testing for the purpose of determining dominant degradation modes and for comparative evaluation of such systems as pressure transmitters is useful. Accelerated testing refers to testing of a system at higher-than-normal stress levels for a short time to estimate system performance following long term exposure to normal stress levels.

Previous studies have shown that temperature generally contributes more to the degradation of pressure transmitters than other stressors. Therefore, the tests in this project were performed at elevated temperatures to induce aging in a short period of time to accommodate the project duration while allowing a preliminary aging study. The Arrhenius equation was used to calculate the approximate age of the transmitters using an activation energy of 0.78 electron volts (eV) which is common for nuclear plant pressure transmitters made by Barton, Foxboro, Rosemount and Tobar. This is the activation energy of the weakest link in the transmitters. The weakest link is in the sensor electronics.

In addition to thermal aging, the transmitters in this project were exposed to larger-than-normal levels of humidity, vibration, pressure cycling, and overpressurization which were applied separately or synergistically when possible. Since the stress factors are often inter-dependent, it is highly desirable to apply all aging environments at once. However, this is often difficult to accomplish in practice.

12. DESCRIPTION OF EXPERIMENTAL SETUP

The experiments reported herein were performed at AMS' laboratory using twenty-two nuclear type pressure transmitters. A listing of these transmitters and their pressure ranges is given in Table 12.1. Note that an identification (I.D.) number, as shown in the last column of Table 12.1, was assigned to each transmitter to facilitate the reporting of the experimental results. The list in Table 12.1 consists of at least four transmitters from each of the common manufacturers of nuclear plant pressure transmitters. The transmitters were obtained from Oak Ridge National Laboratory, Sandia National Laboratories, Arkansas Power and Light Company, and Tobar Incorporated, and a few were purchased for this project. All these transmitters are of the type used in nuclear power plants.

Almost all the equipment and software used in this project were already available from previous projects or related current work except for the aging chambers. Two chambers were used during the project. One was provided by the Mechanical Engineering Department of the University of Tennessee with which AMS has a working agreement and another (a larger chamber) was used through a subcontract with a local company in Knoxville, Tennessee. A picture of one of the environmental chambers is shown in Figure 12.1.

The data acquisition system used in this project consisted of instrumentation amplifiers, electronic filters, strip chart recorders, and microcomputers with built-in analog-to-digital converters for data analysis. A typical data acquisition set up is shown in Figure 12.2. This is followed by a picture and schematic diagram of the laboratory test loop used for performing the noise tests (Figure 12.3). The loop is made of copper tubing and is equipped with a pump to circulate room temperature water at ambient pressure. The loop is designed to provide flow fluctuations for the noise tests. It has a test section with various valves which can be manipulated to simulate sensing line problems and to inject air into the sensing line for testing purposes.

The test equipment for implementing the conventional methods and for a linearity check of the transmitters are found in Section 9. Additional equipment used in this project includes a test unit for performing the power interrupt (PI) test. For the assessment of response time methods, two high-speed reference transmitters made by the Validyne Company were used.

TABLE 12.1

Listing of Pressure Transmitters Used
in This Project

Item	Manufacturer	Model	Pressure Range	Sensor I.D.
1	Barton	752	0-450 inch	PS 19
2	Barton	763	0-1200 psi	PS 18
3	Barton	764	135-44 inch	PS 20
4	Barton	764	0-100 inch	PS 23
5	Foxboro	E11GM/E	0-300 psi	PS 11
6	Foxboro	E11DM/B	0-50 psi	PS 1 L
7	Foxboro	E11DM/B	0-70 psi	PS 1 M
8	Foxboro	E11DM/B	0-20 psi	PS 2 L
9	Foxboro	E11DM/B	0-70 psi	PS 2 M
10	Foxboro	E11DM/B	0-120 psi	PS 2 H
11	Foxboro	E11DM/B	0-20 psi	PS 3
12	Foxboro	E11DM/B	-4-20 psi	PS 4
13	Foxboro	E11GM/E	0-300 psi	PS 8
14	Foxboro	E13DM/H	0-400 inch	PS 9 M
15	Foxboro	E13DM/H	0-550 inch	PS 9 H
16	Foxboro	E13DM/M	0-100 inch	PS 6
17	Foxboro	E13DM/M	0-44.42 inch	PS 5
18	Rosemount	1152GP	0-20 psi	PS 7
19	Rosemount	1153DB	0-550 inch	PS 21
20	Rosemount	1153GD	0-1200 psi	PS 14
21	Rosemount	1154DP	200-0 inch	PS 12
22	Tobar	76DP2	0-135 inch	PS 17
23	Tobar	76DP2	0-550 inch	PS 16
24	Tobar	32PA1	1700-2500 psi	PS 15
25	Tobar	32PA2	0-2000 psi	PS 24
26	Tobar	32DP2	0-250 inch	PS 25

inch = inches of water column
psi = pounds per square inch

Figure 12.1. Photograph of Environmental Chamber
With Pressure Transmitters for Aging
Studies.

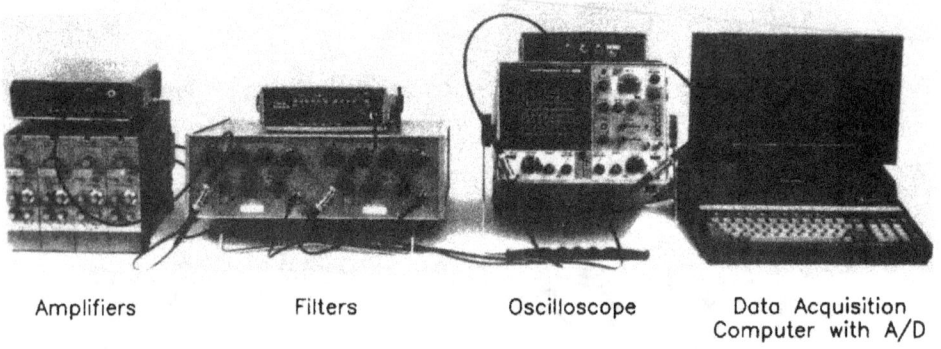

Amplifiers   Filters   Oscilloscope   Data Acquisition
Computer with A/D

Figure 12.2. Arangement of the Data Acquisition System.

Photograph of a Test Section of the Loop

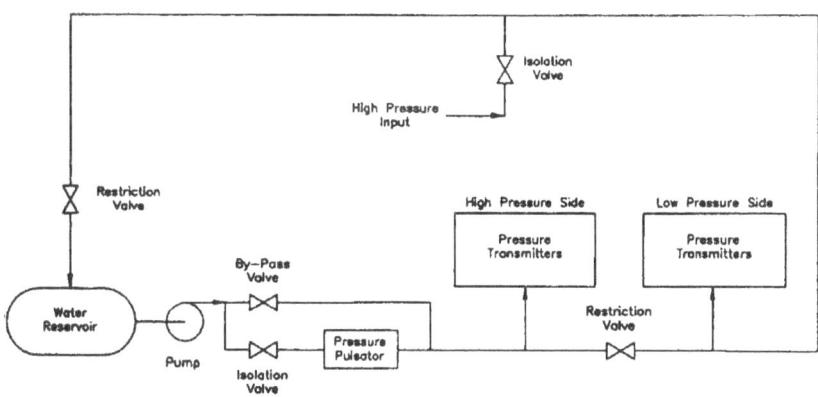

Simplified Schematic Diagram of the Loop

Figure 12.3.   Photograph of Flow Loop and Equipment
for Testing of Pressure Transmitters and
Sensing Lines.

13. ASSESSMENT OF RESPONSE TIME TESTING METHODS

This section presents the results of the experiments performed to evaluate the response time testing methods. The results presented here include a comparison of response time values obtained with each method, repeatability testing of each method, and a discussion of the effect of nonlinear behavior of transmitters on response time.

13.1 COMPARISON OF RESPONSE TIME RESULTS

An experimental assessment of response time testing methods was performed by laboratory testing of all pressure transmitters available. Each transmitter was tested using ramp and step signals, low and high pressure sinusoidal signals, noise analysis, and power interrupt tests as applicable. The results are given in Table 13.1. The ramp test results are used as the reference for evaluating the methods. This is because the ramp test is the most widely used method, and it is important to determine if the other four tests can duplicate the ramp test results. Several points about the results in Table 13.1 are noteworthy:

- Power interrupt (PI) results are given only for force-balance pressure transmitters because the PI method is applicable only to force-balance transmitters.

- Noise analysis results are not given for high pressure transmitters because high pressure noise testing was beyond the scope of this project.

- The ramp, noise, and PI test data were analyzed to provide the ramp time delay (τ_D) of the transmitters, and the step test data were analyzed to give the time constant. The frequency test data were plotted in the form of a Bode diagram. This was used to obtain an approximate response time value corresponding to the reciprocal of the break frequency in the Bode diagram. The fast response transmitters could not be tested by the frequency method because of the dynamic limitation of the frequency test with equipment used in this project.

- The results given for ramp and step tests are based on input signals which increase with time (i.e., positive step signals and increasing ramp signals were used).

The results of the four methods other than the ramp test are within better than about 0.10 seconds of the ramp test results except for a few cases of step and frequency test results. This information and previous experience with these methods indicate that the five methods can be viewed as reasonably close and generally adequate with the step and frequency tests having the lowest agreement with the ramp method. Table 13.2 presents a qualitative evaluation of the five methods. This table includes an overall ranking of the five methods.

TABLE 13.1

Comparison of Response Time Results

Sensor I.D.	Response Time (sec)				
	Ramp	Step	Bode	Noise	PI
PS1 L	0.13	0.12	0.15	0.16	0.14
PS1 M	0.21	0.30	0.32	0.19	0.31
PS2 L	0.11	0.11	0.13	0.15	0.13
PS2 M	0.16	0.20	0.16	0.13	0.23
PS2 H	0.09	0.24	0.12	0.14	0.15
PS3	0.12	0.12	0.16	0.13	0.15
PS4	0.16	0.10	0.09	0.13	0.17
PS5	0.29	0.50	0.42	0.30	0.33
PS6	0.25	0.60	0.23	0.17	0.31
PS7	0.05	0.05	0.05	0.05	N/A
PS9 M	0.28	0.31	0.13	0.24	0.25
PS9 H	0.15	0.23	0.17	0.14	0.17
PS12	0.32	0.38	0.70	0.31	N/A
PS14	<0.01	0.02	N/A	N/A	N/A
PS15	<0.01	0.08	N/A	N/A	N/A
PS16	0.08	0.09	0.10	0.09	N/A
PS17	0.15	0.22	0.35	0.25	N/A
PS18	<0.01	<0.01	N/A	N/A	N/A
PS19	0.05	0.05	0.03	0.11	N/A
PS20	0.17	0.23	0.19	0.18	N/A
PS21	0.07	0.05	0.05	0.09	N/A
PS23	0.17	0.29	0.19	0.22	N/A
PS24	0.08	0.08	N/A	N/A	N/A
PS25	0.33	0.43	0.47	0.35	N/A

Note: 1.   Ramp, noise and PI results correspond to ramp time delay, step results correspond to overall time constant, and Bode results correspond to reciprocal of break frequency.

2.   Above results are nominal response time values obtained from laboratory testing of each transmitter and do not include the corrections to account for test uncertainties.

TABLE 13.2

Qualitative Comparison of Pressure Transmitter
Response Time Testing Methods

Method	Accuracy	Repeatability	Data Analysis	ALARA Problems	Applicability	Overall Ranking
Ramp	High	Moderate	Straight Forward	Yes	All Transmitters	High
Step	Moderate	Low	Straight Forward	Yes	Non-Oscillatory Transmitters	Medium
Frequency	Low	Low	Straight Forward	Yes	Slow Transmitters	Low
Noise	Moderate	Moderate	Requires Computer Analysis	No	All Transmitters	High
PI	High	High	Requires Computer Analysis	No	Force-Balance Transmitters	High

The ranking accounts for all advantages and disadvantages of each method in arriving at an appropriate relative rank. Also included in Table 13.2 is a column to specify ALARA problems with each model. (ALARA stands for As Low As Reasonably Achievable. This is a concept promoted in the nuclear industry to keep the personnel radiation exposure to a minimum.) The tests which require working in the reactor containment may involve exposure to the test personnel and are therefore considered in Table 13.2 as having ALARA problems.

13.2 PRESENTATION OF RAW DATA FOR EACH METHOD

For each of the five methods, a typical raw data transient is shown in Figure 13.1 to 13.5. The traces for the ramp and step tests each contain two pairs of strip chart traces showing both normal and oscillatory tests. When the output traces are free of oscillations, the calculation of ramp time delay or time constant is straight forward because the traces can be used directly to identify the response time. With the ramp test, for example, the delay between the output of the transmitter under test and the high speed reference transmitter is identified as the response time of the transmitter being tested. For the step tests, the time to reach 63.2 percent of final value of the output is identified as the transmitter time constant.

When the output traces are oscillatory, direct calculation of the ramp time delay or time constant is difficult and depends on the magnitude of the oscillations. When oscillations are encountered, one must first verify that the oscillations are not due to improper equipment operation or test setup. If the oscillations are unavoidable, the tests have to be done in a manner to allow response time calculation after the oscillations have died out or by performing transfer function analysis. Practical considerations can sometimes make it very difficult to completed a test and calculate a reliable response time in the presence of oscillations.

The frequency test results are equal to the reciprocal of the break frequency in the Bode plots as illustrated in Figure 13.3. This calculation assumes that the transmitter is a first order system and therefore the results of the Bode plots are approximate values unless the Bode plot is mathematically fitted to an appropriate function.

For noise analysis and PI tests, the data were sampled with a digital data acquisition system and then analyzed with appropriate software. A typical PSD plot from noise analysis for one of the transmitters is shown in Figure 13.6.

There are cases where the PI or noise tests can be complicated. For example, the PI data can contain an overshoot as shown in Figure 13.7. The overshoot is not unique to a particular transmitter model but tends to occur more often with low pressure transmitters.

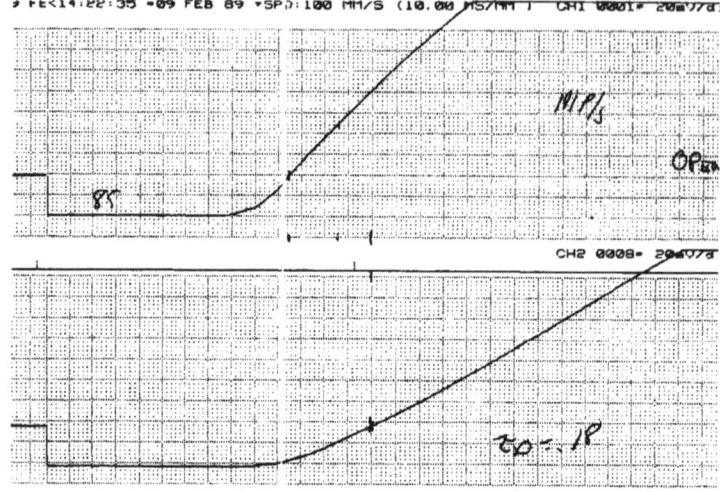

a) Normal Case

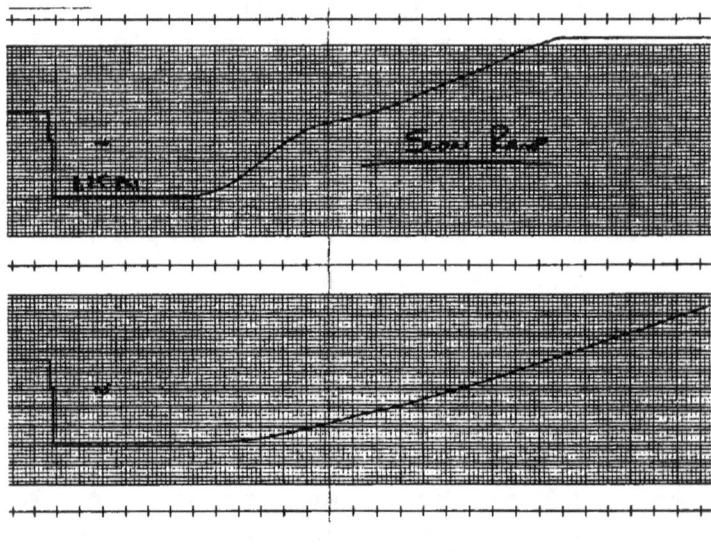

b) Oscillatory Case

Figure 13.1. Ramp Test Transients for a Normal
Test (a) and a Test Where Oscillations
were Encountered (b).

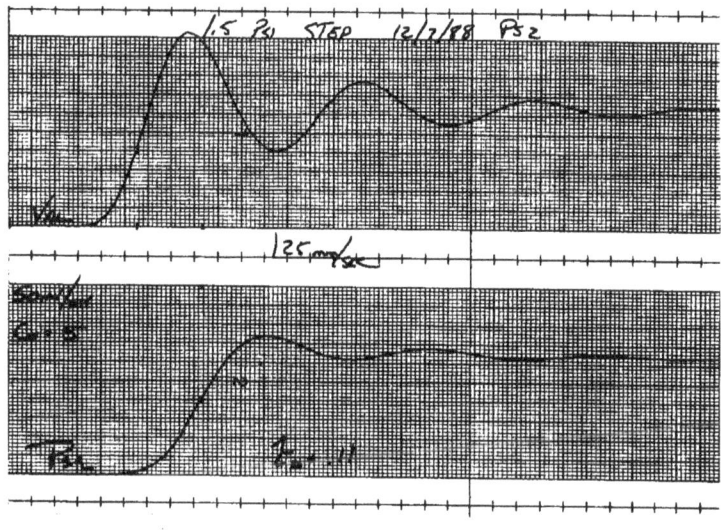

a) Oscillatory Case

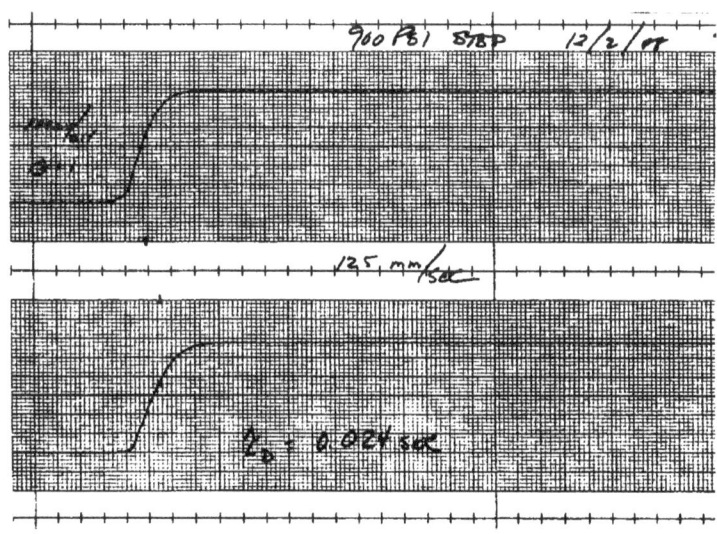

b) Normal Case

Figure 13.2.   Step Response Traces for an
               Oscillatory Pressure Transmitter
               (a) and a Normal Pressure
               Transmitter (b).

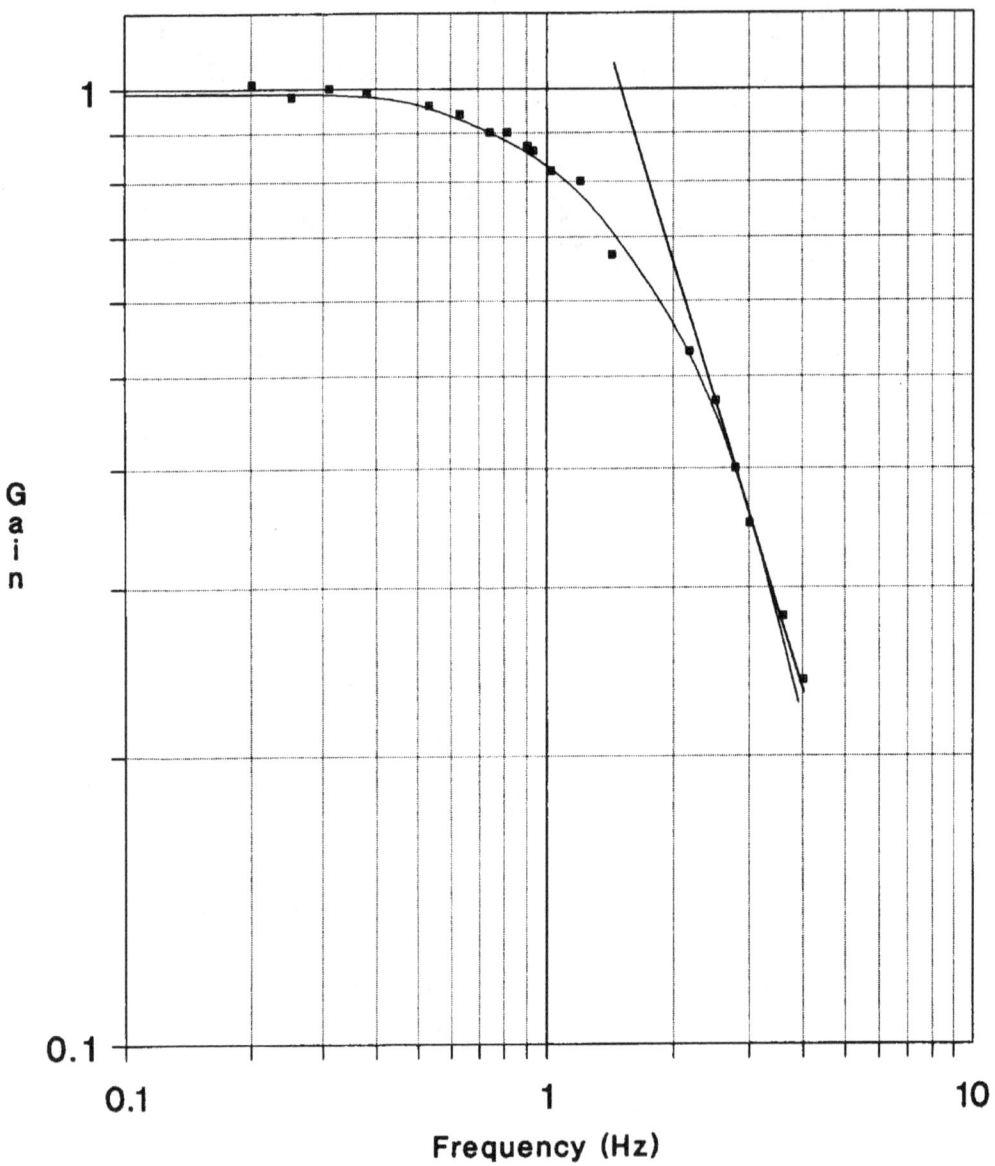

Figure 13.3. Bode Plot from the Frequency Test
of a Pressure Transmitter.

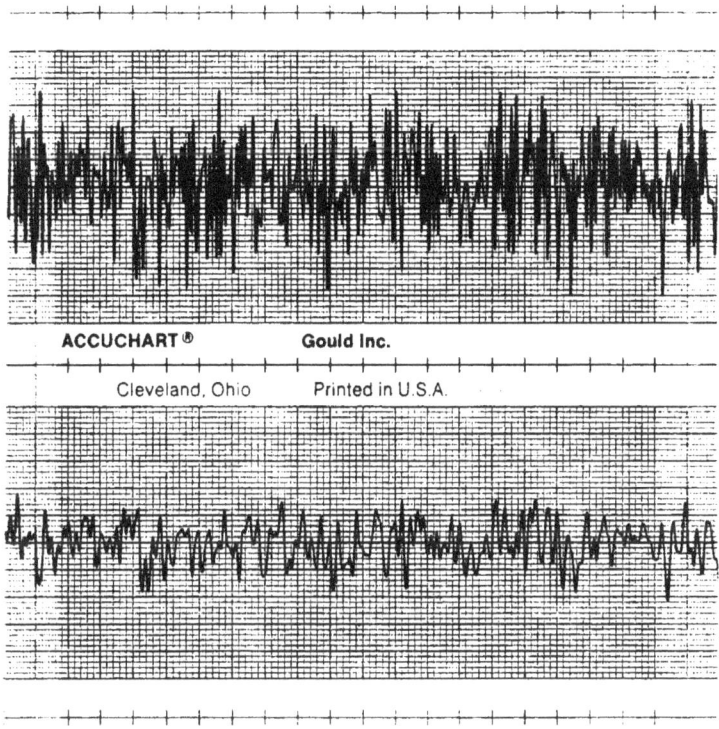

ACCUCHART®   Gould Inc.

Cleveland, Ohio   Printed in U.S.A.

Figure 13.4.   Noise Data Traces for a
Reference Transmitter (top)
and a Transmitter Under
Test (bottom).

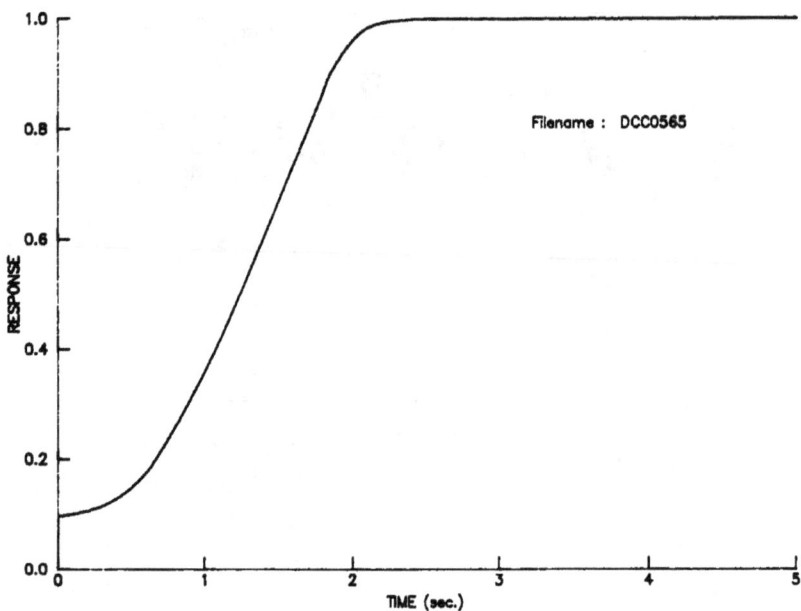

Figure 13.5.   Power Interupt Transient.

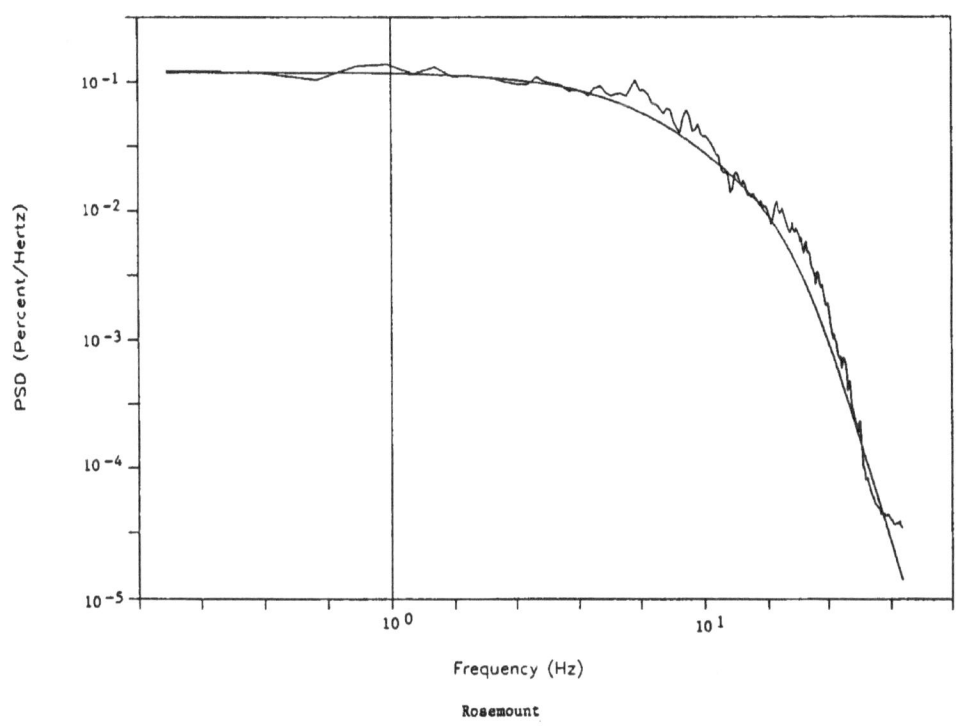

Figure 13.6. A Typical PSD and the FIT to the PSD from
Laboratory Testing of a Pressure Transmitter.

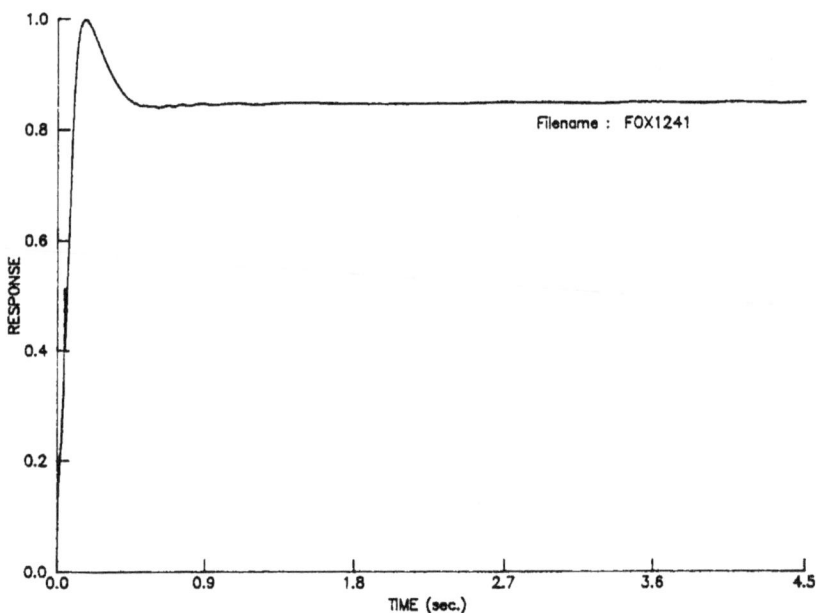

Figure 13.7.   PI Data of a Transmitter With Overshoot.

13.3   REPEATABILITY OF RESPONSE TIME TESTING METHODS

The repeatability of the five test methods was examined by performing repeated tests on a number of pressure transmitters. The results are discussed below.

Repeatability of Ramp Test. The repeatability of ramp test can be affected by pressure measurement tolerances due to hysteresis and other effects in pressure transmitters. The significance of this problem was investigated by testing of representative pressure transmitters with various ramp rates. The results are given in Table 13.3. These results demonstrate that there is not a significant dependence between response time and ramp rate as long as reasonable ramp rates are used in testing the transmitters. Table 13.4 shows the representative results of repeated ramp tests using nominal ramp rates for each sensor. It is apparent that the repeatability of the ramp test results is generally good. The maximum difference in the results of the repeated ramp tests (as shown in the right-hand column of Table 13.4) is less than 0.10 seconds except for one case.

Repeatability of Step Test. The repeatability of step tests was investigated by performing repeated tests with various step sizes. The results are tabulated in Table 13.5. The repeatability of the step test results for a few of the transmitters depends on the magnitude of the step, but in most cases the repeatability is reasonable. The comparison between the step test and ramp test results shows close agreement for a majority of the transmitters. Table 13.6 compares nominal ramp and step test results for a selected number of transmitters. These results indicate that, the step test results are often slightly higher than ramp test results. This conclusion is supported by equation 8.10 and 8.11 that show that time constant is generally larger than ramp time delay. The step response results given here are equal to the time required for the sensor output to reach 63.2 percent of its final value following a step change in input pressure.

Repeatability of Frequency Test. The repeatability of frequency test results are not as good as the repeatability of the ramp or step test results. This conclusion is based on repeated frequency testing of four pressure transmitters. The results are not tabulated here.

Repeatability of Noise Analysis. Noise analysis repeated on several transmitters over a six month period showed good repeatability in the results. Table 13.7 presents the results of repeated response time tests of sixteen transmitters. The difference between maximum and minimum results of repeated tests are given in the right-hand column. The results are repeatable to better than 0.10 seconds in most cases. This is comparable with the repeatability of the ramp test.

Repeatability of PI Test. The repeatability of PI test was investigated by repeated testing of several Foxboro pressure transmitters. The results are shown in Table 13.8. The difference between the maximum and minimum results of repeated tests is also shown. The results are repeatable to better than 0.10 seconds.

TABLE 13.3

Response Time Versus Ramp Rate Results

Sensor I.D.	Ramp Rate [†] (psi/sec)	Response Time [†] (sec)	Difference (sec)
PS 1 L	5, 10, 14, 17	0.13, 0.13, 0.14, 0.13	0.01
PS 1 M	5, 11, 19	0.23, 0.21, 0.20	0.03
PS 2 L	2.5, 5, 8, 10	0.10, 0.13, 0.11, 0.10	0.03
PS 2 M	10, 14, 18	0.14, 0.18, 0.17	0.04
PS 2 H	8, 16, 27, 37	0.08, 0.08, 0.09, 0.09	0.01
PS 3	2, 4, 7, 8	0.13, 0.12, 0.12, 0.12	0.01
PS 5 *	0.7, 6, 10	0.28, 0.30, 0.30	0.02
PS 7	1, 2, 4, 11, 25, 30	0.04, 0.04, 0.05, 0.05, 0.04, 0.04	0.01
PS 8	20, 33, 63, 100 115, 136, 150, 180	0.14, 0.21, 0.21, 0.18 0.18, 0.21, 0.17, 0.18	0.07
PS 9 M	2, 5	0.28, 0.23	0.05
PS 9 H	6, 9, 11	0.15, 0.15, 0.14	0.01
PS 11	30, 33, 60, 63, 83, 100, 115, 136	0.17, 0.19, 0.19, 0.20, 0.19, 0.19, 0.20, 0.20	0.03
PS 12	1, 2, 5, 6	0.32, 0.30, 0.30, 0.30	0.02

Continued on next page

TABLE 13.3 (continued)

Sensor I.D.	Ramp Rate [†] (psi/sec)	Response Time [†] (sec)	Difference (sec)
PS 14	62, 130, 190, 250	<0.01, <0.01, <0.01, <0.01	0.00
PS 15	140, 400, 520	<0.01, <0.01, <0.01	0.00
PS 16	5, 20, 29, 30	0.10. 0.07, 0.07, 0.07	0.03
PS 17	3, 6, 7, 9, 10	0.20, 0.16, 0.15, 0.13, 0.12	0.08
PS 18	250, 700, 900, 238 666, 833	<0.01, <0.01, <0.01, <0.01 <0.01, <0.01	0.00
PS 19	1, 4	0.04, 0.05	0.01
PS 20	1, 2	0.17, 0.17	0.00
PS 21	2, 7, 10, 11	0.05, 0.07, 0.08, 0.08	0.03
PS 23	1, 3, 4	0.20, 0.17, 0.17	0.03
PS24	155, 307, 317 455, 571	0.07, 0.07, 0.07 0.08, 0.08	0.01
PS25	1, 2, 5, 6	0.39, 0.39, 0.27, 0.27	0.12

*   *Ramp Rates for this sensor are in terms of inches of water per second.*

†   *The ramp rates and response times listed in each row correspond to one another respectively, e.g., ramp rates of 5, 10, 14, and 17 in the first row of this table correspond to the response times 0.13, 0.13, 0.14 and 0.13 respectively.*

    Note: The response time results in this table are for increasing ramp signals.

TABLE 13.4

Repeatability of Ramp Test Results

Sensor I.D.	Date of Test	Test Engineer	Response Time (sec)	Difference (sec)
PS1 L	11/08/88	MH	0.15, 0.16	0.04
	12/13/88	REF	0.13, 0.13, 0.13, 0.13	
	12/15/88	REF	0.14, 0.12, 0.15, 0.13	
PS1 M	11/08/88	MH	0.23, 0.22	0.04
	12/16/88	REF	0.23, 0.21, 0.20, 0.20	
	12/27/88	REF	0.19, 0.19, 0.19, 0.19	
PS2 L	10/17/88	MH	0.18, 0.16, 0.16	0.08
	12/07/88	REF	0.10, 0.13, 0.11, 0.10	
	12/08/88	REF	0.12, 0.12, 0.13, 0.11	
PS2 M	03/13/89	REF	0.14, 0.18, 0.17	0.04
PS2 H	03/13/89	REF	0.08, 0.08, 0.09, 0.09	0.01
PS3	10/18/88	MH	0.16, 0.16, 0.16	0.04
	11/22/88	REF	0.12, 0.14, 0.14, 0.12	
	12/01/88	REF	0.14, 0.12, 0.12, 0.12	
	12/02/88	REF	0.13, 0.12, 0.12	
PS4	10/06/88	MH	0.16, 0.16, 0.16	0.00
PS5	10/19/88	MH	0.28, 0.28, 0.30, 0.30	0.02
PS6	10/25/88	MH	0.24, 0.26, 0.24	0.02
PS7	10/06/88	MH	0.04, 0.04, 0.04	0.01
	12/08/88	REF	0.05, 0.05, 0.04, 0.04	
	12/09/88	REF	0.05, 0.05, 0.04, 0.04	
PS8	01/20/89	REF	0.21, 0.21, 0.21, 0.21	0.07
	02/09/89	REF	0.14, 0.18, 0.17, 0.18	
PS9 M	10/06/88	MH	0.28, 0.28, 0.28, 0.23	0.05
PS9 H	03/13/89	REF	0.15, 0.15, 0.14	0.01

Continued on next page

TABLE 13.4 (continued)

Sensor I.D.	Date of Test	Test Engineer	Response Time (sec)	Difference (sec)
PS11	01/19/89	REF	0.19, 0.20, 0.20, 0.20	0.03
	02/09/89	REF	0.17, 0.19, 0.19, 0.19	
PS12	10/07/88	MH	0.32, 0.32, 0.32	0.03
	02/14/89	REF	0.29, 0.30, 0.32	
	02/15/89	REF	0.29, 0.30, 0.31	
PS14	12/02/88	REF	<0.01, <0.01, <0.01	N/A
	12/01/88	REF	<0.01, <0.01, <0.01	
PS15	02/13/89	REF	<0.01, <0.01, <0.01	N/A
PS16	01/20/89	REF	0.10, 0.07, 0.07, 0.07	0.03
	02/09/89	REF	0.10, 0.08, 0.08. 0.07	
PS17	12/06/88	REF	0.20, 0.16, 0.15, 0.13, 0.12	0.09
	02/09/89	REF	0.19, 0.18, 0.13, 0.11	
PS18	02/09/89	REF	<0.01, <0.01, <0.01	N/A
	02/09/89	REF	<0.01, <0.01, <0.01	
PS19	02/01/89	REF	0.04, 0.03, 0.04, 0.05, 0.05,	0.02
	02/02/89	REF	0.05, 0.04, 0.05	
PS20	02/01/89	REF	0.17, 0.16, 0.17, 0.18, 0.18, 0.18, 0.17, 0.18	0.03
	02/02/89	REF	0.15, 0.16, 0.17, 0.16	
PS21	01/31/89	REF	0.06, 0.07, 0.08, 0.08	0.03
	02/02/89	REF	0.05, 0.06, 0.07, 0.07	
PS23	03/04/89	REF	0.17, 0.17, 0.17, 0.20 0.17, 0.16, 0.17, 0.21	0.05
PS 24	04/07/89	REF	0.07, 0.07, 0.07, 0.07 0.08, 0.08, 0.08, 0.08	0.01
PS25	04/07/89	REF	0.39, 0.40, 0.38, 0.39 0.27, 0.32, 0.27, 0.29	0.13

Note: 1. The results in this table are from tests with nominal ramp rates.
2. MH and REF are the initials of test personnel who performed the repeatability tests.

TABLE 13.5

Repeatability of Step Test Results

Sensor I.D.	Applied Step (psi)	Response Time (sec)
PS1 low	1, 2, 5, 8	0.14, 0.13, 0.10, 0.09
PS1 med	10, 20, 35, 70	0.28, 0.32, 0.38, 0.61
PS2 low	0.5, 1, 1.5, 2 2.5, 3	0.13, 0.12, 0.11, 0.11 0.10, 0.09
PS2 med	10, 20, 50, 70	0.17, 0.20, 0.35, 0.50
PS2 high	20, 50, 80, 120	0.24, 0.22, 0.26, 0.45
PS3	1, 1.5, 2, 2.5	0.16, 0.12, 0.12, 0.12
PS4	1, 2, 3, 4	0.11, 0.10, 0.09, 0.08
PS5	0.3, 0.5, 0.7, 1, 1.5	0.39, 0.50, 0.62, 0.71, 0.91
PS6	0.5, 1, 1.5, 2	0.64, 0.80, 0.94, 1.23
PS7	2, 4, 6, 8, 10, 12	0.05, 0.05, 0.05, 0.04 0.04, 0.05,
PS8	40, 80, 120, 150 200, 250	0.34, 0.56, 0.72, 0.78 0.78, 0.86
PS9 med	2, 4, 6, 10 12.5, 15	0.28, 0.30, 0.31, 0.39 0.45, 0.45
PS9 high	5, 10, 15, 20	0.18, 0.23, 0.32, 0.44
PS11	40, 80, 120, 150 200, 250	0.35, 0.50, 0.61, 0.68 0.67, 0.72
PS12	1, 1.5, 2, 3, 5	0.40, 0.31, 0.40, 0.39, 0.38
PS14	100, 400, 700, 900	0.02, 0.02, 0.02, 0.02

continued on next page

TABLE 13.5 (continued)

Sensor I.D.	Applied Step (psi)	Response Time (sec)
PS15	500, 1000, 1500	0.07, 0.08, 0.09
PS16	2, 4, 8, 10 12, 15	0.11, 0.10, 0.09, 0.08 0.08, 0.09
PS17	0.5, 1, 1.5, 2 2.5, 3, 3.5, 4 4.5, 5	0.22, 0.22, 0.27, 0.30 0.30, 0.32, 0.32, 0.32 0.34, 0.32
PS18	200, 400, 600, 800 1000, 1200	<0.01, <0.01, <0.01, <0.01 <0.01, <0.01
PS19	1, 2, 5, 10, 15	0.05, 0.05, 0.05, 0.05, 0.05
PS20	0.25, 0.50, 1, 2, 3	0.24, 0.23, 0.23, 0.24, 0.23
PS21	1, 3, 5, 10, 15, 20	0.06, 0.03, 0.02, 0.05, 0.06, 0.07
PS23	0.5, 1, 1.5, 2, 3, 4	0.28, 0.30, 0.29, 0.29, 0.29, 0.29
PS24	500, 1000, 1500	0.07, 0.08, 0.08
PS25	1, 3, 5, 8	0.45, 0.44, 0.40, 0.42

The applied step signals are increasing steps.

TABLE 13.6

Comparison of Response Time Results
From Ramp Test and Step Test

Sensor I.D.	Response Time (sec) Ramp	Step	Response Time Difference (sec)
PS 1	0.21	0.30	0.09
PS 2	0.11	0.11	0.00
PS 3	0.12	0.12	0.00
PS 4	0.16	0.10	-0.06
PS 7	0.05	0.05	0.00
PS 9	0.28	0.31	0.03
PS 12	0.32	0.38	0.06
PS 16	0.08	0.09	0.01
PS 17	0.15	0.22	0.07
PS 19	0.05	0.05	0.00
PS 20	0.17	0.23	0.06
PS 21	0.07	0.05	-0.02
PS 23	0.17	0.29	0.12
PS 24	0.08	0.08	0.00
PS 25	0.33	0.43	0.10

The results are from tests with increasing ramp and positive step signals with nominal ramp rates and step sizes.

TABLE 13.7

Repeatability of Noise Analysis Results

Sensor I.D.	Date of Test	Measured Response Time (sec)	Difference (sec)
PS1 low	01/07/89	011, 0.12	0.06
	01/28/89	0.16, 0.16	
	01/29/89	0.17, 0.16	
	01/31/89	0.17, 0.17	
PS1 med	01/07/89	0.16, 0.16	0.07
	01/28/89	0.21, 0.23	
	01/29/89	0.16, 0.16	
	01/31/89	0.17, 0.17	
PS2 low	01/04/89	0.15	0.02
	01/28/89	0.14, 0.14, 0.17, 0.14	
PS2 med	03/25/89	0.13, 0.13	0.00
PS3	01/04/89	0.13, 0.13	0.02
	01/07/89	0.13, 0.13	
	01/28/89	0.14, 0.12, 0.12, 0.14, 0.14	
PS5	01/07/89	0.32, 0.27, 0.28, 0.28 0.28	0.10
	01/28/89	0.23, 0.34, 0.33	
	02/09/89	0.24	
PS7	12/08/88	0.05, 0.05, 0.06	0.03
	09/30/88	0.03, 0.04, 0.06, 0.04	
PS8	01/29/89	0.07	0.01
	01/31/89	0.07, 0.08, 0.07	
PS9 med	01/07/89	0.21, 0.19, 0.21, 0.21	0.05
	01/28/89	0.22, 0.22, 0.24, 0.24	

continued on next page

TABLE 13.7 (continued)

Sensor I.D.	Date of Test	Measured Response Time (sec)	Difference (sec)
PS11	01/29/89	0.09	0.02
	01/31/89	0.07	
PS17	01/03/89	0.26	0.06
	01/06/89	0.20, 0.25	
	01/28/89	0.26, 0.25, 0.26	
PS19	02/11/89	0.10, 0.11, 0.12, 0.11	0.02
PS20	02/11/89	0.17, 0.17, 0.18, 0.18	0.01
PS21	02/11/89	0.09, 0.09, 0.10, 0.08, 0.08, 0.10	0.02
PS23	04/04/89	0.23, 0.22, 0.22, 0.22	0.01
PS25	05/05/89	0.33, 0.35, 0.36, 0.38	0.05

TABLE 13.8

Repeatability of PI Test Results

Sensor I.D.	Date of Test	Response Time (sec)	Difference (sec)
PS1 low	10/24/88	0.15, 0.15, 0.15, 0.15	0.06
	12/13/88	0.12, 0.13, 0.10, 0.14	
		0.15, 0.16	
PS1 med	10/24/88	0.26, 0.27, 0.30, 0.28	0.07
	12/16/88	0.33, 0.31, 0.28, 0.30	
PS2 low	10/24/88	0.11, 0.09, 0.10, 0.14	0.06
	12/07/88	0.11, 0.09, 0.08, 0.11	
		0.12, 0.13, 0.13, 0.14	
		0.14, 0.14, 0.14	
PS2 med	03/13/89	0.30, 0.24, 0.23, 0.22, 0.21	0.09
PS2 high	03/13/89	0.16, 0.15, 0.15, 0.15	0.01
PS3	10/24/88	0.13, 0.11, 0.13, 0.15	0.05
	11/22/88	0.12, 0.11, 0.11, 0.15	
		0.15, 0.16, 0.16, 0.16	
PS4	10/24/88	0.17, 0.17, 0.16, 0.16	0.01
PS5	10/25/88	0.33, 0.32, 0.35, 0.33	0.03
PS6	10/25/88	0.31, 0.32, 0.31, 0.32	0.01
PS9 med	10/25/88	0.24, 0.27, 0.26, 0.25	0.03
PS9 high	03/13/89	0.19, 0.17, 0.17, 0.17, 0.17	0.02

13.4 EFFECT OF LINEARITY ON RESPONSE TIME

The response time result with any of the five methods is strongly affected by the linearity of the pressure transmitter being tested. Therefore, the linearity characteristics of the transmitter used in this project were qualitatively examined. The examination involved using the pressure waveform generator shown earlier in Section 10 to provide test signals with varying amplitudes and frequencies. The signals were simultaneously applied to the transmitter under test and to a reference transmitter. The output of both transmitters was recorded for qualitative evidence of nonlinearities. These tests revealed two pressure transmitters with nonlinear behavior. These transmitters were kept in the project and tested along with the other transmitters. In almost all tests, the results for these two transmitters were in notable disagreement with all other test results.

Figure 13.8 shows linearity test traces for a normal and a nonlinear transmitter. The signal pair on the top is from the reference transmitter and a linear transmitter. The signal pair on the bottom is from the reference transmitter and a nonlinear transmitter. Note that the nonlinearity of the transmitter is manifested as an asymmetrical sinewave output as seen in the last trace on the bottom of Figure 13.8.

Further investigation of nonlinearity of the two transmitters involved ramp testing with increasing and decreasing ramp signals. So far in this report, all results reported for ramp tests have been based on increasing ramps. If transmitters are linear, the results will not be affected significantly by the direction of the applied test signal. Therefore, increasing ramp signals are usually used for practical reasons. Table 13.9 shows the results of the ramp tests for the two nonlinear transmitters in comparison with four linear transmitters. Results are given for both increasing and decreasing ramp signals. Note that the response times of the nonlinear transmitters are significantly dependent on the direction of the test signal while the results from the linear transmitters are essentially the same regardless of the characteristics of the input test signal.

The cause of the nonlinearity of these two transmitters is not known. These two transmitters are from a manufacturer whose other transmitters have no linearity problem. These nonlinear transmitters are gage pressure transmitters.

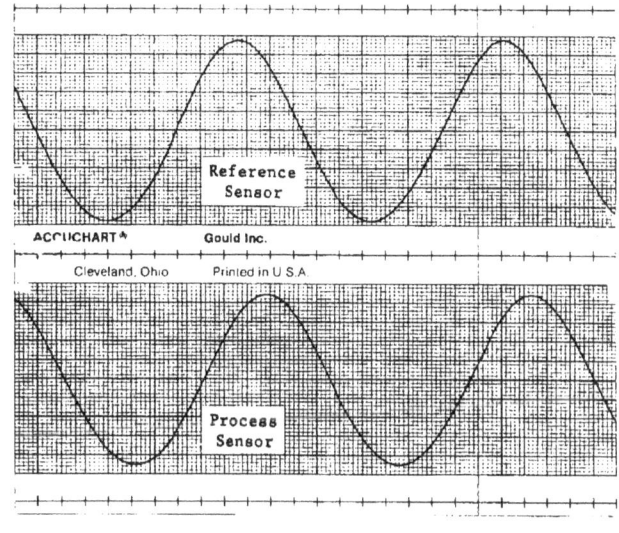

Linear

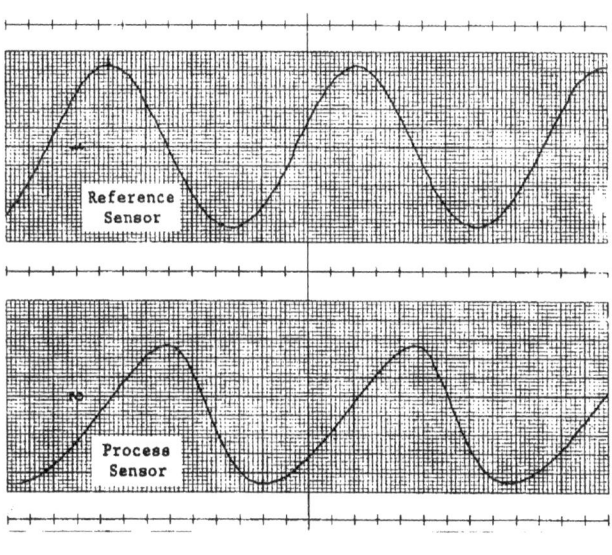

Non-linear

Figure 13.8. Strip Chart Traces Showing Linear and Non-linear Transmitters.

TABLE 13.9

Effect of Ramp Direction on Linear
Versus Nonlinear Pressure Transmitters

Sensor Linearity	Response Time (sec)	
	Up Ramp	Down Ramp
Nonlinear	0.21	0.03
Nonlinear	0.20	0.03
Linear	0.13	0.14
Linear	0.16	0.12
Linear	0.05	0.05
Linear	0.17	0.22
Linear	0.07	0.06

The above ramp test results are from increasing (up) and decreasing (down) test signals with nominal ramp rates.

14. RESPONSE TIME VERSUS CALIBRATION RESULTS

It was discussed in Section 9 that changes in response time of pressure transmitters will not necessarily be accompanied by changes in their calibration and vice versa. This was demonstrated experimentally using a Rosemount Model 1152 pressure transmitter which has a damping potentiometer. The transmitter was calibrated and its response time was first measured with the damping potentiometer set at its minimum (zero damping). The damping potentiometer was then turned about 1/3 of the way toward its maximum setting (about 33% damping), and the calibration and response time tests were repeated. The results indicated negligible changes in calibration, but the response time increased by about an order of magnitude (Table 14.1). In Table 14.2 we have shown the results of an experiment where the zero and span (i.e., calibration) was changed to determine its effect on response time. The results indicate no change in response time in spite of significant changes in calibration.

The response time results in Table 14.1 are given for both ramp and step tests with different ramp rates and step sizes. It is clear that changes in response time by as much as a factor of ten can occur without a significant change in calibration. Opposite situations can be envisioned where large calibration shifts can occur without affecting the intrinsic response time of the transmitter. For example, zero shifts can occur with no effect on response time.

The Rosemount Model 1152 transmitter was also tested with the noise method. Figure 14.1 shows the raw data with and without damping. Note that the effect of damping is manifested by a large decrease in the amplitude of the transmitter's output fluctuations.

TABLE 14.1

Response Time Changes
as a Function of Damping

Ramp Test

Ramp Rate (psi/sec)	Response Time (sec)	
	No Damping	33% Damping
5	0.04	0.40
7	0.05	0.34
15	0.04	0.29

Step Test

Size of Step (psi)	Response Time (sec)	
	No Damping	33% Damping
5	0.05	0.56
10	0.05	0.58
15	0.04	0.56

TABLE 14.2

Effect of Calibration Changes on Response Time of a Pressure Transmitter

Zero Changes	Span Changes	Response Time (sec)
-10%	0	0.10
-20%	0	0.11
+10%	0	0.11
+7%	1%	0.11
+30%	5%	0.11
-33%	-4%	0.11

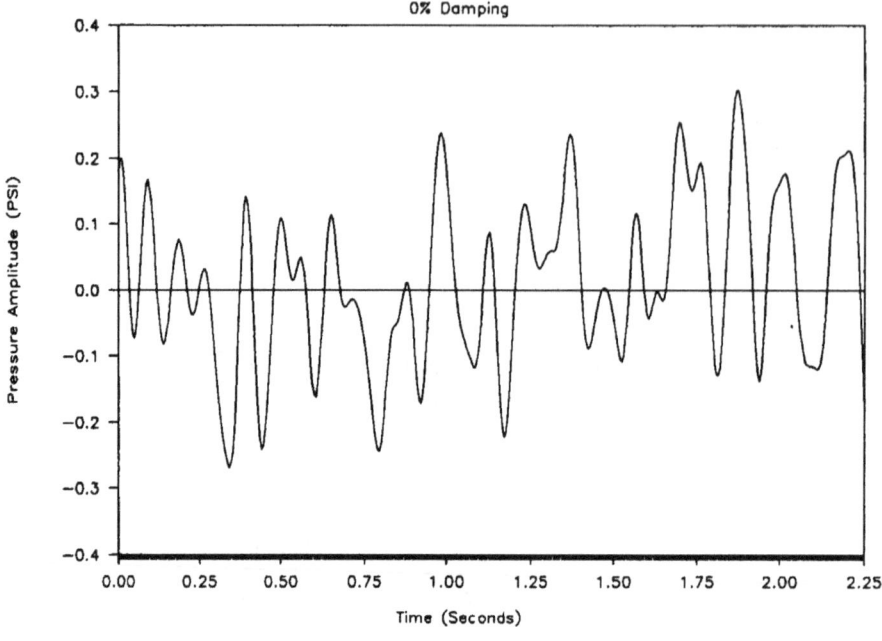

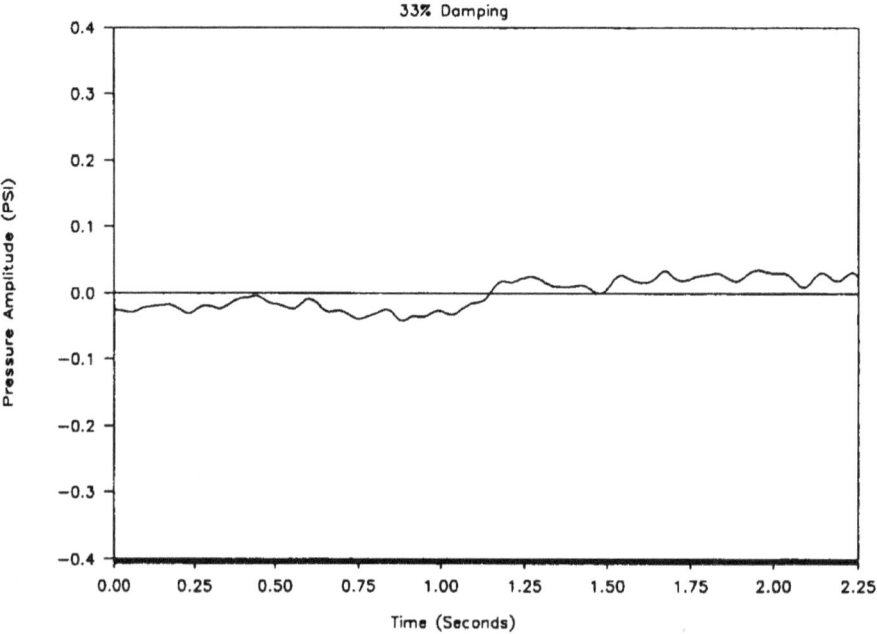

Figure 14.1.   Output Fluctuations of a Pressure
Transmitter With and Without Damping.

15. EXPERIMENTS WITH SENSING LINES

To demonstrate the effect of sensing line blockages and voids, a few experiments were conducted in the laboratory both using noise analysis and the ramp test. These experiments involved injecting air and simulating blockages in laboratory sensing lines using hand operated valves. Figure 15.1 shows the test section of the laboratory loop illustrating the valves that can be manipulated to simulate sensing line problems. The noise method was first used to study sensing line problems. Figure 15.2 shows the noise test results. Pressure fluctuations are shown in this figure for a Rosemount flow transmitter with and without an air bubble in its housing. It is apparent that air has acted as a filter to remove some of the high frequency components of the noise signal. The air bubble also caused the sensor/sensing line system to have a larger response time. Figure 15.3 compares the PSDs of the transmitter with and without air. Note that the PSD rolls off at a notably smaller frequency for the case when air exists in the system. That is, the response time is larger with air in the system.

In addition to causing an increase in response time, this air in the system caused a resonance which is apparent in Figure 15.3 in the PSD of the signal for the transmitter with air. The resonance peak is located at a frequency of about 3 Hz. Monitoring of resonance peaks on PSD plots due to air or blockage in the sensing lines is an effective tool for detecting of sensing line problems in nuclear power plants. The resonance peaks due to air usually move to lower frequencies as the size of the air bubble increases. In the case of the blockages, width of the resonance increases as the blockage is increased.

Another experiment with sensing line effects performed in this project involved a test in which noise data were obtained with the low side of the sensing line blocked in one case and the high side blocked in another case. The raw data are shown in Figure 15.4 for a Tobar differential pressure transmitter used for this experiment. The figure consists of three plots as follows:

- Both sides of the transmitter open to main flow line. This corresponds to a normal case.

- The high-side of the transmitter is closed and the low-side is open to the main flow line.

- The low-side of the transmitter is closed, but the high-side is open to the main flow line.

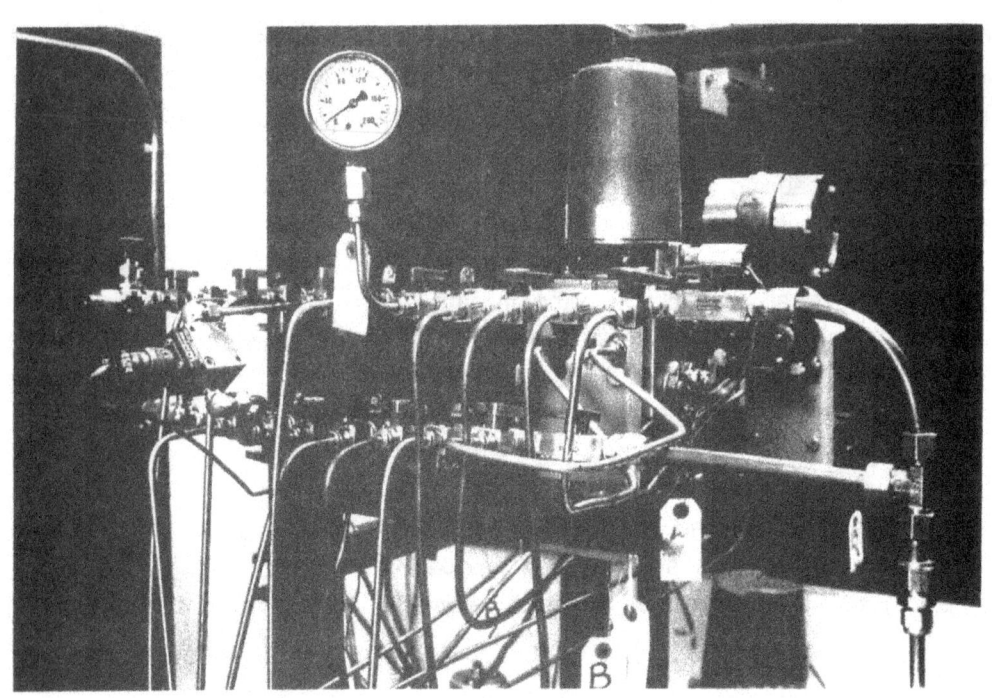

Figure 15.1.   Photograph of Test Section of the
Laboratory Flow Loop.

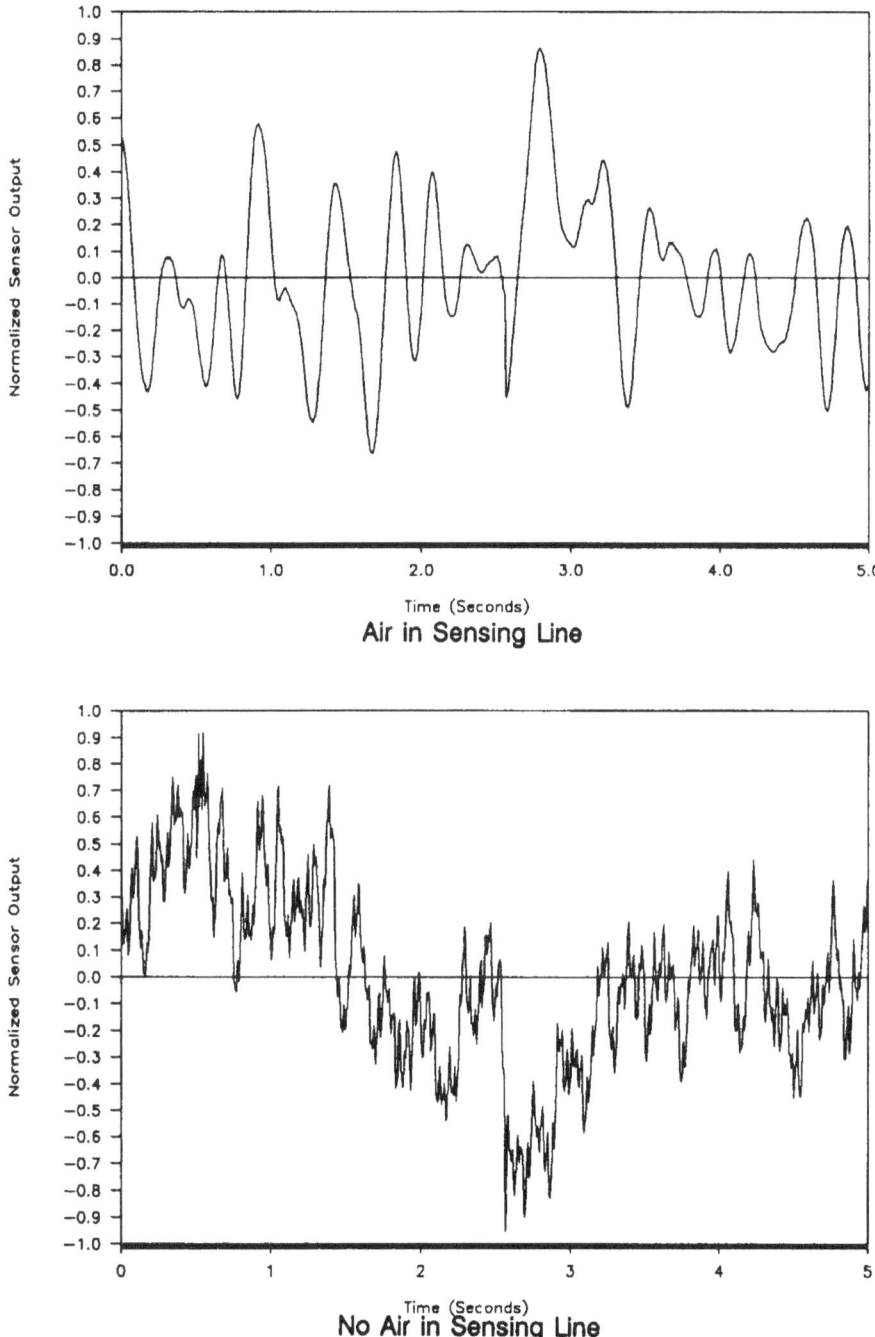

Figure 15.2. Typical Pressure Transmitter Noise Data
With and Without Air in the Sensing Line.

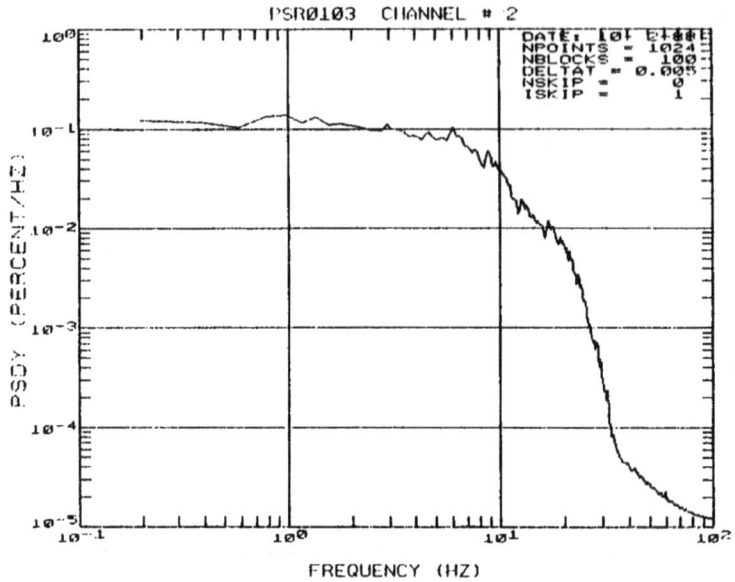

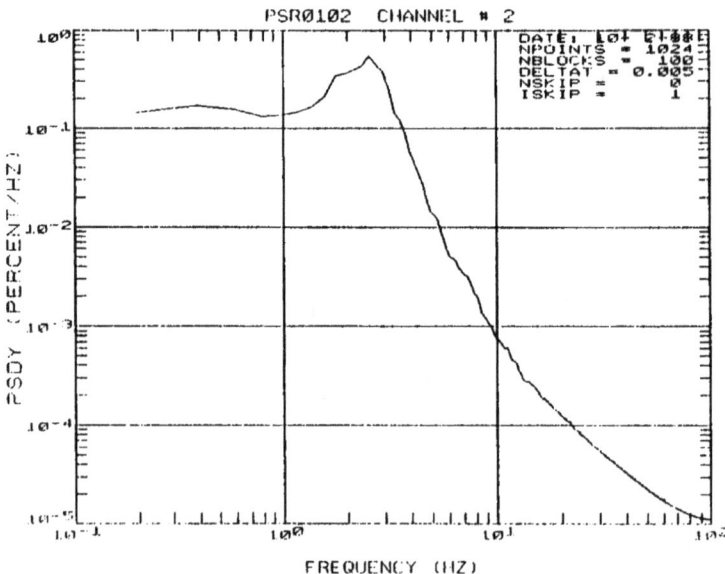

Figure 15.3. PSDs of a Transmitter Without (top)
and With (bottom) Air in the System.

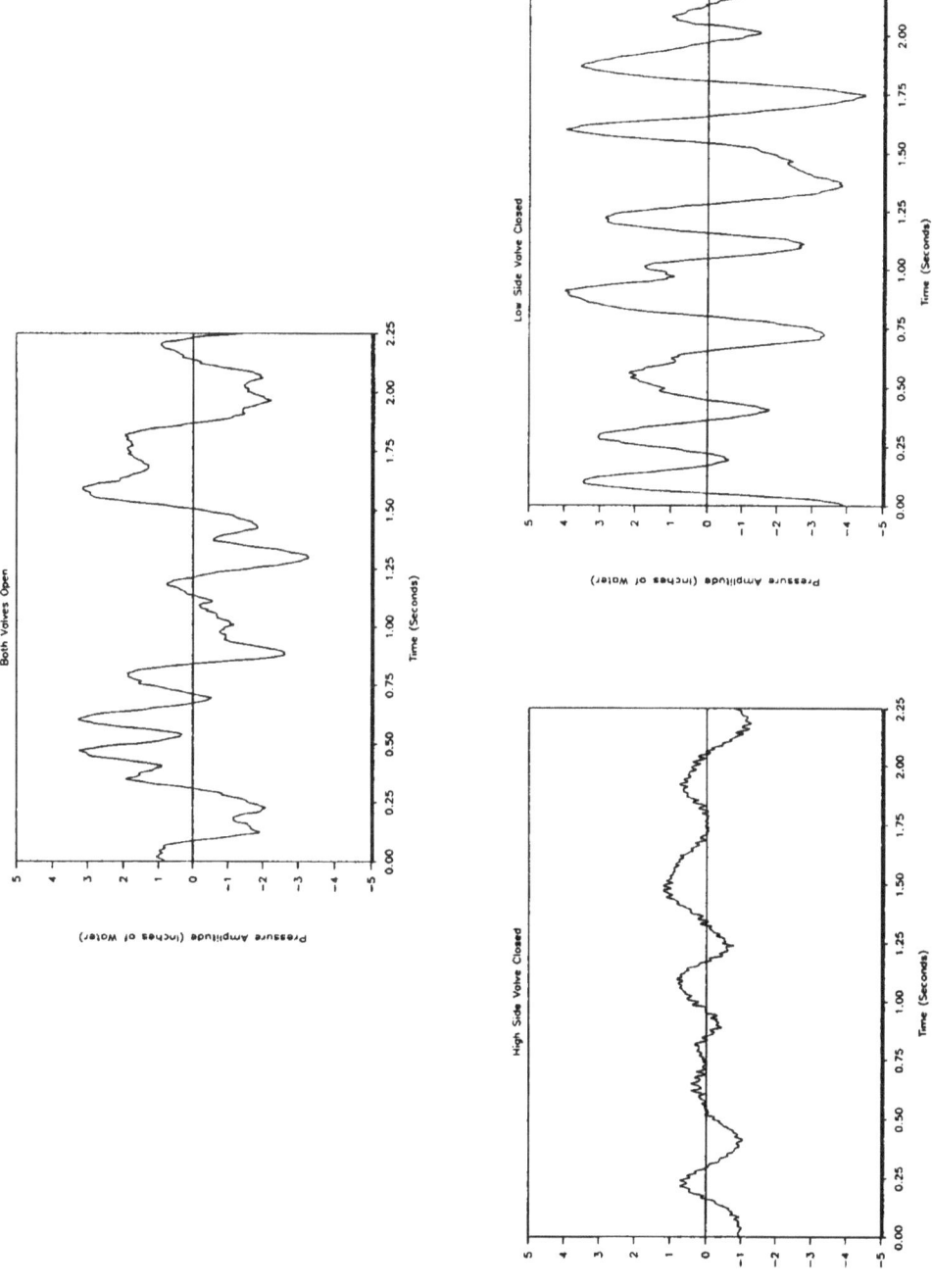

Figure 15.4 Noise Output of a Tobar Differential Pressure Sensor With Sensing Lines Open (top), High Side Blocked (left), and Low Side Blocked (right).

Note that in one case (high side valve closed), the amplitude of the fluctuations decreased significantly with the blockage. A comparison of the plot on top with the two plots on the bottom of Figure 15.4 shows that the dynamic characteristics of the fluctuations are different with a blockage in either sensing line. This information is useful in determining blockages or leaks in sensing lines.

The effect of air in the sensing line was also studied using the ramp method. This work involved testing of a Foxboro pressure transmitter with the hydraulic ramp generator. Tests were performed with various volumes of air in the line between the ramp generator and the transmitter. The results are given in Table 15.1.

TABLE 15.1

Response Time as a Function of
Air in Sensing Line

Length of Bubble (inch)	Response Time (sec)
0	0.12
8	0.13
20	0.16
30	0.19
60	0.39

1. *Above tests were performed with a 25-foot, 1/4-inch O.D. Nylon Tubing between the test unit and the transmitter under test.*

2. *The bubble lengths given above are approximate values.*

16. RESULTS OF AGING RESEARCH

An experimental aging study was initiated and preliminary aging results were obtained for several transmitters after exposure to heat, humidity, vibration, pressure cycling, and overpressurization. These conditions were applied individually or combined when possible. In all cases, higher-than-normal levels of stress were applied to accommodate accelerated testing.

Both the response time and calibration of the transmitters were identified before and after aging. A data acquisition system was developed and used to monitor the output of the transmitter during the aging process to identify any gross malfunction or failure, if encountered. The results are discussed below.

Thermal Aging and Pressure Cycling. A number of transmitters were installed in environmental chambers where they were exposed to temperatures of 165 to 195°F combined with pressure cycling. The cycling involved pressure step signals with pressure amplitudes corresponding to the calibrated range of the transmitters. The pressure cycling was induced using a signal generator to activate a solenoid valves to apply and vent pressure every few seconds. The results are given in Table 16.1 in terms of the percentage change in calibration and response time of the transmitters after aging. The number of pressure cycles and the estimated age of the transmitters are listed for each transmitter. Note that the estimated thermal age and the number of pressure cycles are not the same for all transmitters because they were aged at different rates.

Assuming that pressure transmitters in nuclear power plants are normally exposed to a temperature of 120°F, the equivalent age of the transmitters at the chamber temperatures was estimated using the Arrhenius equation. An activation energy of 0.78 eV was used in estimating the accelerated thermal age of the transmitters.

The results as shown in Table 16.1 indicate some zero shifts and span changes for most of the transmitters. For four cases, the zero shifts are large even though they are not accompanied by large span changes. In the case of response time, although minor degradations are shown by the results, a definite conclusion cannot be made on whether or not measurable degradations occurred during this aging process. Note that the changes of less than 1% for calibration or less than 10% for response time are considered negligible.

TABLE 16.1

Thermal Aging and Pressure Cycling Results

Sensor I.D.	Induced Aging		Performance Changes		
	Thermal (years)	# Cycles	Zero	Span	Response Time
PS1	2.2	130,000	20%	16%	15%
PS2	3.0	170,000	180%	<1%	18%
PS3	0.25	26,000	<1%	<1%	<10%
PS4	2.2	127,000	69%	14%	20%
PS7	1.7	61,000	2%	2%	<10%
PS8	0.6	N/A	2%	<1%	14%
PS9	2.2	127,000	16%	<1%	14%
PS12	0.6	2,200	17%	2%	18%
PS14	0.6	N/A	8%	5%	<10%
PS16	0.6	2,200	11%	<1%	25%
PS18	0.6	N/A	131%	19%	<10%
PS19	0.6	2,200	12%	8%	<10%
PS20	0.6	2,200	<1%	<1%	<10%
PS21	1.9	110,000	<1%	3%	15%
PS23	0.6	2,200	34%	6%	<10%
PS24	0.6	N/A	5%	<1%	<10%
PS25	0.6	2,200	<1%	<1%	15%

Heat and Humidity. Humidity was introduced in a number of transmitters by moisturizing the inside wall of the housing caps that cover the electronics. The transmitters were then installed in an environmental chamber at 195°F for a few days to induce an equivalent of three months of thermal aging. The results are given in Table 16.2 in terms of percentage change in calibration and response time due to this aging environment. The output of the transmitters was monitored by a data acquisition system while the transmitters were in the environmental chamber. One transmitter failed after a few days in the chamber. This transmitter was kept in the chamber along with the others and was examined at the end of the aging process. This examination revealed that the span potentiometer of the transmitter had failed. However, the response time of the transmitter could be tested after recalibration to a different pressure range and was found to have experienced little degradation. The remaining transmitters showed negligible changes in calibration and response time, but the potentiometers in several of these transmitters had erratic behavior after this aging process.

Vibration and Cycling Results. A simple shaker table was built for this project (Figure 16.1) and used for vibration testing. The vibration of interest in this project was that of normal operational vibration as opposed to seismic vibration. Several transmitters were tested for performance degradation due to combined effects of vibration and pressure cycling. The combination of vibration and pressure cycling was selected because this was simple to accommodate. The results of vibration aging experiments are given in Table 16.3 in terms of percent changes in calibration and response time characteristics. The vibration intervals and the number of imposed cycles are also listed in the table. The cycling was performed by imposing step pressure signals in the calibrated range of the transmitter every few seconds. The results showed minor changes in performance for all transmitters.

Overpressurization. Pressure transmitters experience overpressurizations during maintenance, calibration and reactor trips. This condition was simulated in the laboratory using numerous overpressurization cycles. The effects of overpressurization on calibration and response time are shown in Table 16.4. The number of pressure step signals and their magnitudes are also listed along with the normal calibration range of the transmitters. The effect of overpressurization produced minor changes in calibration and response time, with the latter being the more pronounced degradation.

TABLE 16.2

Results of Humidity Tests

| Sensor I.D. | Performance Changes | | |
	Zero	Span	Response Time
PS3	<1%	<1%	<10%
PS7	4%	<1%	<10%
PS9	<1%	<1%	21%
PS12	2%	2%	<10%
PS19	<1%	<1%	<10%
PS16	4%	<1%	<10%
PS23	*	*	12%
PS25	<1%	<1%	11%

* Sensor could not be calibrated due to failure of potentiometer early in the aging process.

Figure 16.1.   Vibration Aging Setup.

TABLE 16.3

Vibration and Pressure Cycling Results

Sensor I.D.	Time at 0.3G	# Cycles	Performance Changes		
			Zero	Span	Response Time
PS21	26 Hrs.	5500	<1%	<1%	17%
PS4	24 Hrs.	5400	8%	<1%	<10%
PS1	7 Hrs.	1300	<1%	<1%	<10%
PS16	10 Hrs.	1600	4%	<1%	11%
PS7	8 Hrs.	2000	<1%	<1%	<10%
PS25	7 Hrs.	1700	<1%	<1%	10%
PS19	8 Hrs.	2000	4%	2%	<10%
PS3	8 Hrs.	2000	<1%	<1%	<10%
PS23	24 Hrs.	5400	<1%	<1%	<10%
PS9	15 Hrs.	3600	2%	<1%	<10%
PS12	25 Hrs.	5600	<1%	<1%	<10%
PS20	25 Hrs.	5600	<1%	<1%	10%
PS17	26 Hrs.	5500	<1%	<1%	20%

TABLE 16.4

Results Of Overpressurization Tests

Sensor I.D.	Calibration Range	# Cycles	Step (psi)	Performance Changes		
				Zero	Span	Response Time
PS25	0-250"wc	1600	80	3%	4%	22%
PS12	200-0"wc	1600	80	11%	2%	<10%
PS17	0-135"wc	1600	80	7%	2%	22%
PS23	0-100"wc	1600	80	<1%	<1%	17%
PS9	0-400"wc	2200	80	2%	<1%	11%
PS4	-4-20psi	2200	80	6%	<1%	<10%
PS21	0-550"wc	2200	80	4%	2%	20%
PS19	0-450"wc	2200	80	3%	3%	20%
PS3	0-20psi	4500	80	<1%	2%	<10%
PS7	0-20psi	4500	80	<1%	<1%	<10%
PS16	0-550"wc	4500	80	<1%	3%	12%
PS20	135-44"wc	4500	80	<1%	<1%	<10%

17. ASSESSMENT OF TYPICAL PLANT PROCEDURES

A majority of commercial nuclear power plants use the ramp method for response time testing of pressure transmitters. A simplified drawing of the test equipment (called hydraulic ramp generator) for performing the ramp test is shown in Figure 17.1. To perform the test, the pressures in the two accumulators are adjusted so that the pressure in one accumulator is above the setpoint at which the response time is to be measured and the pressure in the other accumulator is below the setpoint. When the "Signal Initiate Solenoid Valve" is opened, the pressures in the two accumulators equalize. This equalization pressure transient is applied to the process transmitter and to a reference transmitter and the outputs are recorded. The response time of the process transmitter is then identified by measuring the time delay between the two transmitters at a particular setpoint pressure. Because of inherent empirical difficulties in performing these tests, it is important that the test procedures be accurate. In order to study the accuracy of typical plant procedures, we reviewed four plant procedures used by different utilities. We determined that these plant procedures are basically adequate and made the following observations or recommendations for better tests.

Process Transmitter Venting. Ramp tests are typically performed at the process transmitter location. The transmitter is first isolated from the process and the hydraulic ramp generator is connected to the transmitter. Although venting or bleeding of a transmitter is very important in preparation to perform the tests, several of the response time procedures we reviewed did not specifically address venting or bleeding. After the hydraulic ramp generator is connected to the process transmitter, the system should be properly vented under pressure to ensure that no air remains trapped inside the process transmitter or the tubing connecting it to the hydraulic ramp generator. Even a small amount of air can have an effect on the measured response time. Instructions and precautions for proper venting should be included in all plant procedures.

Identification of Set Point Pressure. One or more pressure set-points are identified for each transmitter to be tested. The setpoints usually correspond to the safety system trip setpoints. Two methods are used during the actual ramp test to identify the setpoint at which the response time of the transmitter is measured. In the first method, a calibrated gage connected to the hydraulic ramp generator is used to adjust the output pressure to the setpoint. Then the output signals from both the process transmitter and the reference transmitter are recorded on a strip chart recorder. These traces are then labeled on the chart as the setpoints (Figure 17.2). The time response is then measured as the difference between the times at which transmitters pass through their respective setpoint lines after the ramp signal has been applied. Since the setpoints are referenced to a single common measurement (the calibrated pressure gage), any effects of the calibrations of the reference or the process transmitter are eliminated. This provides the process transmitter's "intrinsic response time".

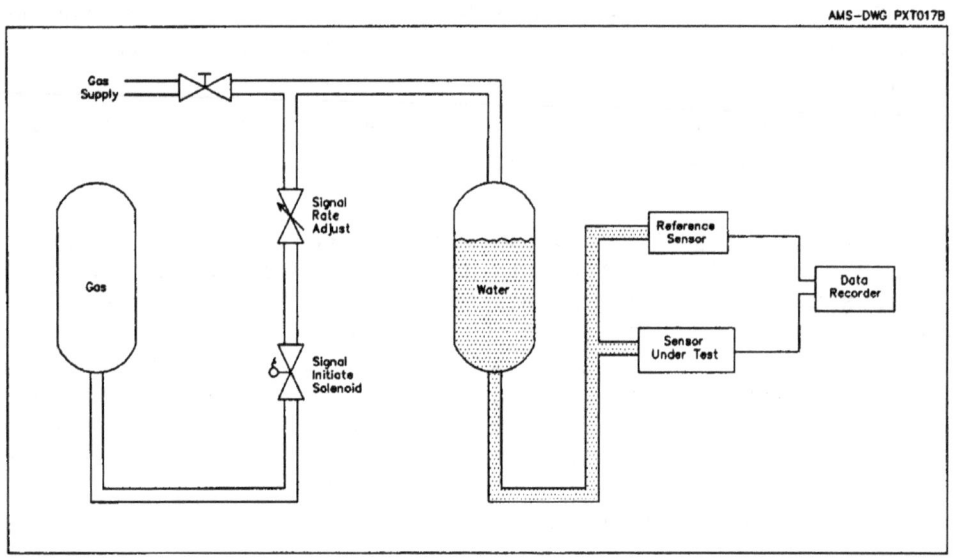

Figure 17.1.   Illustration of the Hydraulic Ramp Generator.

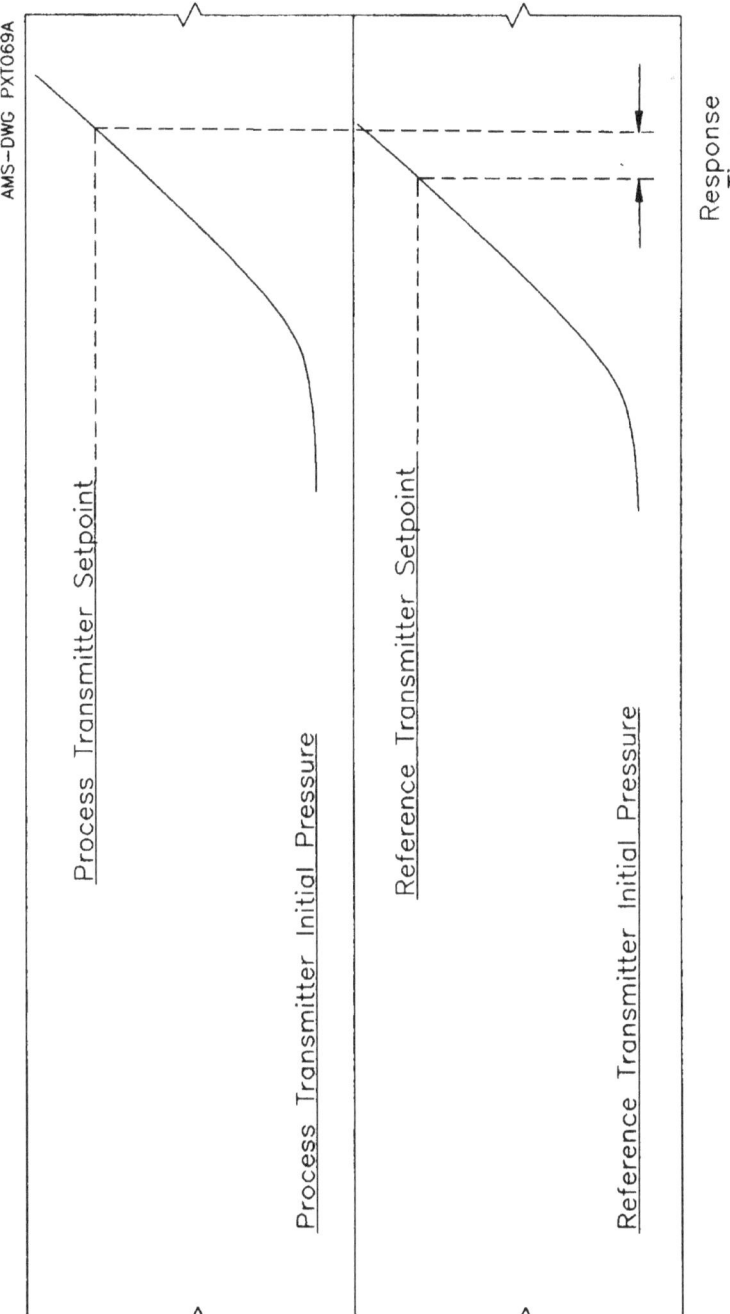

Figure 17.2. Illustration of the Ramp Test Data and Calculation of Response Time.

The second method for setpoint identification is based on the calibrations of the process and reference transmitters. The desired setpoint pressure is converted to equivalent electrical outputs from the transmitters. The calculated electrical values of the setpoints are then identified on the strip chart trace either by supplying the signal with a power supply or based on the strip chart recorder's calibration. These signals are then used to identify the points in time at which the reference and process transmitters indicate the setpoint. With this method, errors in the calibrations of the process or the reference transmitter may result in errors in the measured response time. These errors may result in conservative or non-conservative response time measurements.

Ramp Rate Selection. The pressure measurement tolerance due to hysteresis and other effects in pressure transmitters can contribute to errors in measured response time. Figure 17.3 shows that the response time error caused by pressure measurement tolerance is related to the ramp rate by:

$$Response\ Time\ Tolerance\ (\pm)\ =\ \frac{Reproducibility\ Tolerance\ (\pm\ \%\ of\ span)}{Ramp\ Rate\ (\%\ of\ span/sec.)}$$

For typical manufacturer's specifications of repeatability, accuracy, and hysteresis, the response time uncertainty may be high if low ramp rates are used for performing the tests. The remedy is to use fast ramp signals, if this is allowed. Tests performed in this project, however, have not identified any significant dependence of response time on ramp rate as long as reasonable ramp rates are used.

Total Channel Response Time. Figure 17.4 shows a simplified block diagram of a typical safety channel in a nuclear power plant. All components of the safety channel are periodically tested for time response. There are, however, variations from plant to plant with respect to which of these components are tested together. In some plants, the individual components (i.e., transmitter, electronics, actuation), are tested separately and the individual response times are summed for a total channel response time. This is usually preferred because it allows identification of the process transmitter's response time.

Filtering of Signals. In one procedure, the use of a 5 Hz low-pass filter on the signals was suggested. Sometimes, the measured response times for the pressure transmitters are in the 50 to 100 millisecond range. A 5 Hz low-pass filter has a time constant of about 30 milliseconds and could therefore have an impact on the test results. The use of filters on the output signals during pressure transmitter response time testing should be avoided.

Oscillatory Signals. Oscillatory test outputs are sometimes encountered during the response time tests. These outputs can result from the transmitter under test, the reference transmitter, or the test signal. When these oscillations are observed, the first

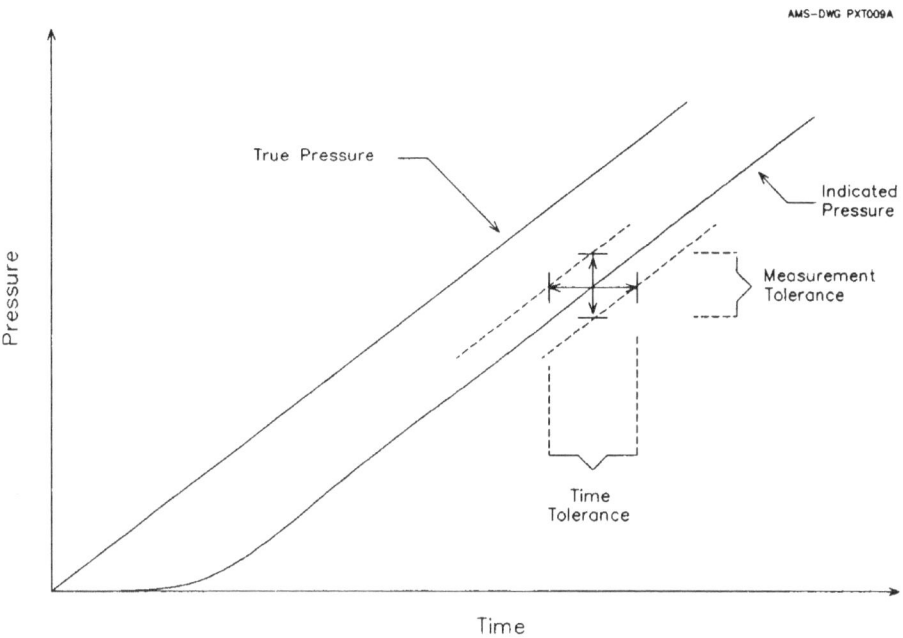

Figure 17.3.   Effect of Pressure Measurement Tolerance
on Response Time Results.

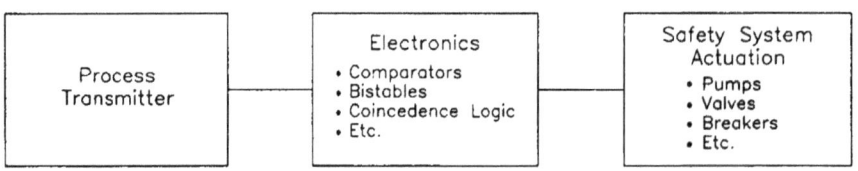

Figure 17.4.   Block Diagram of a Simplified Safety Channel.

step should be to verify that there is no air in the system because air can cause oscillations. If there is no air in the system and the oscillations are not due to problems in the test equipment or set up, the test results must be interpreted carefully to yield reliable results. This is done by making response time calculations from the portion of the output after the oscillations have died out. However, practical considerations sometimes prevent the test personnel from being able to observe the oscillations. In fact, these problems have caused some plants to record apparent negative response times that are encountered if the oscillations are large enough to overshoot the input (Figure 17.5). The current plant procedures do not have a reference to this potential problem.

Acceptance Criteria. Plant procedures often lack a clear acceptance criteria and unambiguous instructions as what steps should be taken if a transmitter fails response time testing. Such procedures could benefit from a statement of the range of acceptable response times and what constitutes an acceptable result. In addition, the accuracy of the test method and the uncertainty of the test results should be accounted for in specifying test methods or acceptance criteria.

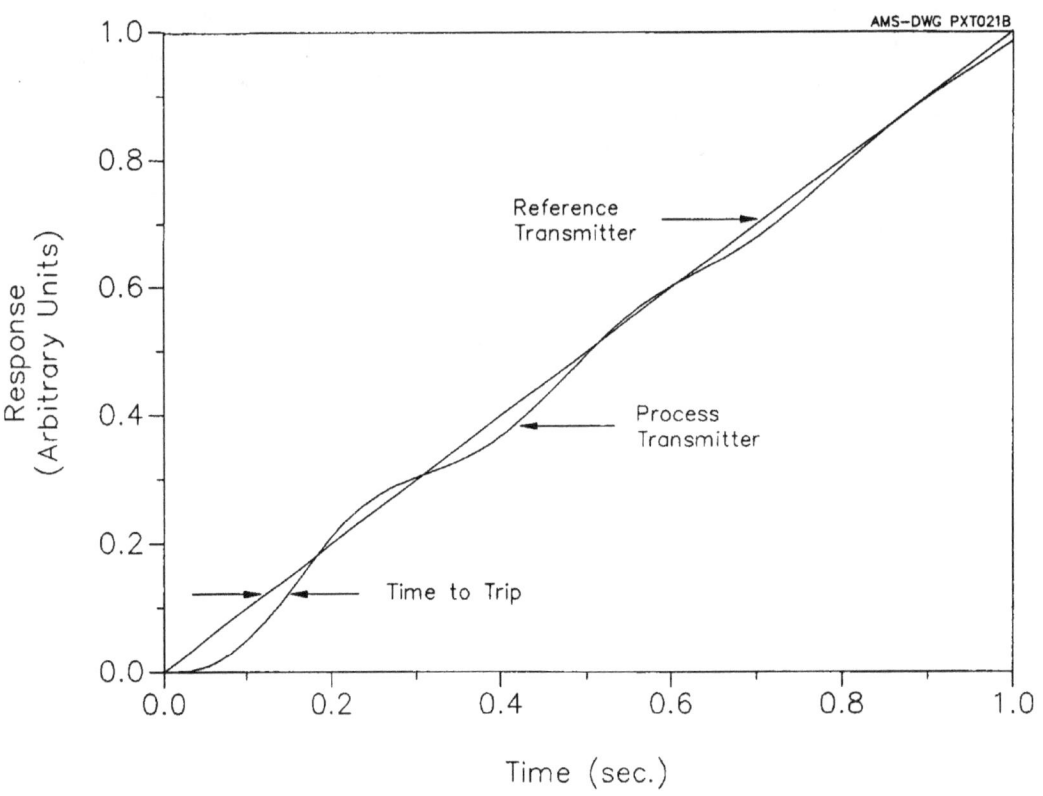

Figure 17.5.   Ramp Response of an Oscillatory Transmitter.

18. SEARCH OF LER DATA BASE

A formal search of the LER data base on reported problems with pressure sensing systems in nuclear power plants was completed in this project. The results are documented in a separate report given in Appendix A. The key points are discussed below.

The search of the LER data base covered the period beginning with 1980 through October 1988. There are about 30,000 LERs in the data base for this period. Pressure transmitter problems in this period were found in 1,325 LERs. This amounts to about 4.4% of all LER (Figure 18.1). The LERs on pressure transmitters are categorized in three groups: personnel related, potential age-related, and others or unknown as shown in Figure 18.1. The percentages in Figure 18.1 add up to 104% because a few LERs are common among the three groups. Figure 18.2 illustrates each of the three groups and gives the total of all groups. Note in Figure 18.2 that there is a notable change in the number of LERs after 1984 when the reporting requirements changed.

Potential age-related problems accounted for 38% of the reported problems, a majority of which affected the calibration of the transmitters. A listing of failures or degradation which were categorized here as potential age-related is given in Appendix A.

Sensing line problems were mentioned in 401 LERs (Figure 18.3). About 27% of the sensing line problems are considered as potential age-related problems. This is second to personnel related problems which consumed 60% of the LERs associated with sensing lines.

Typical LER abstracts describing pressure transmitter response time problems are given in Table 18.1. This is followed by Table 18.2 with typical abstracts of problems involving sensing lines.

A search of other data bases such as the NPRDS and NPE was also considered. The search of the NPRDS data base was not possible in time to be presented here. The NPE data base was studied for pressure transmitter problems and was determined to contain essentially the same information as the LER data base.

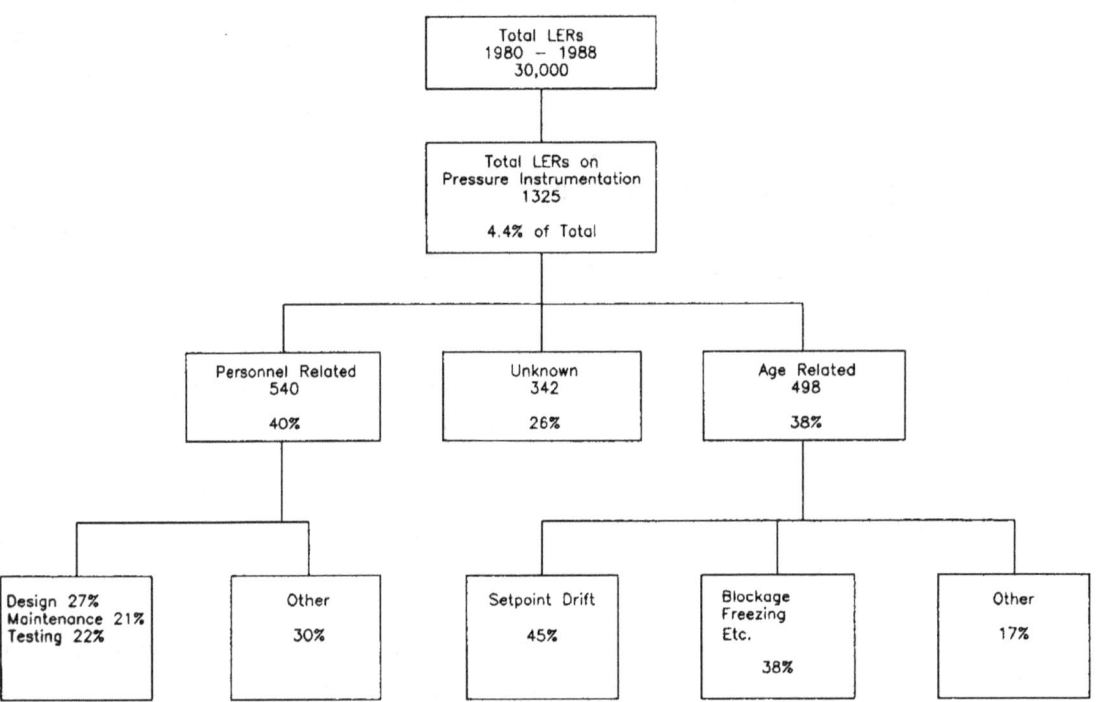

Figure 18.1.   Breakdown of LERs on Pressure Transmitters.

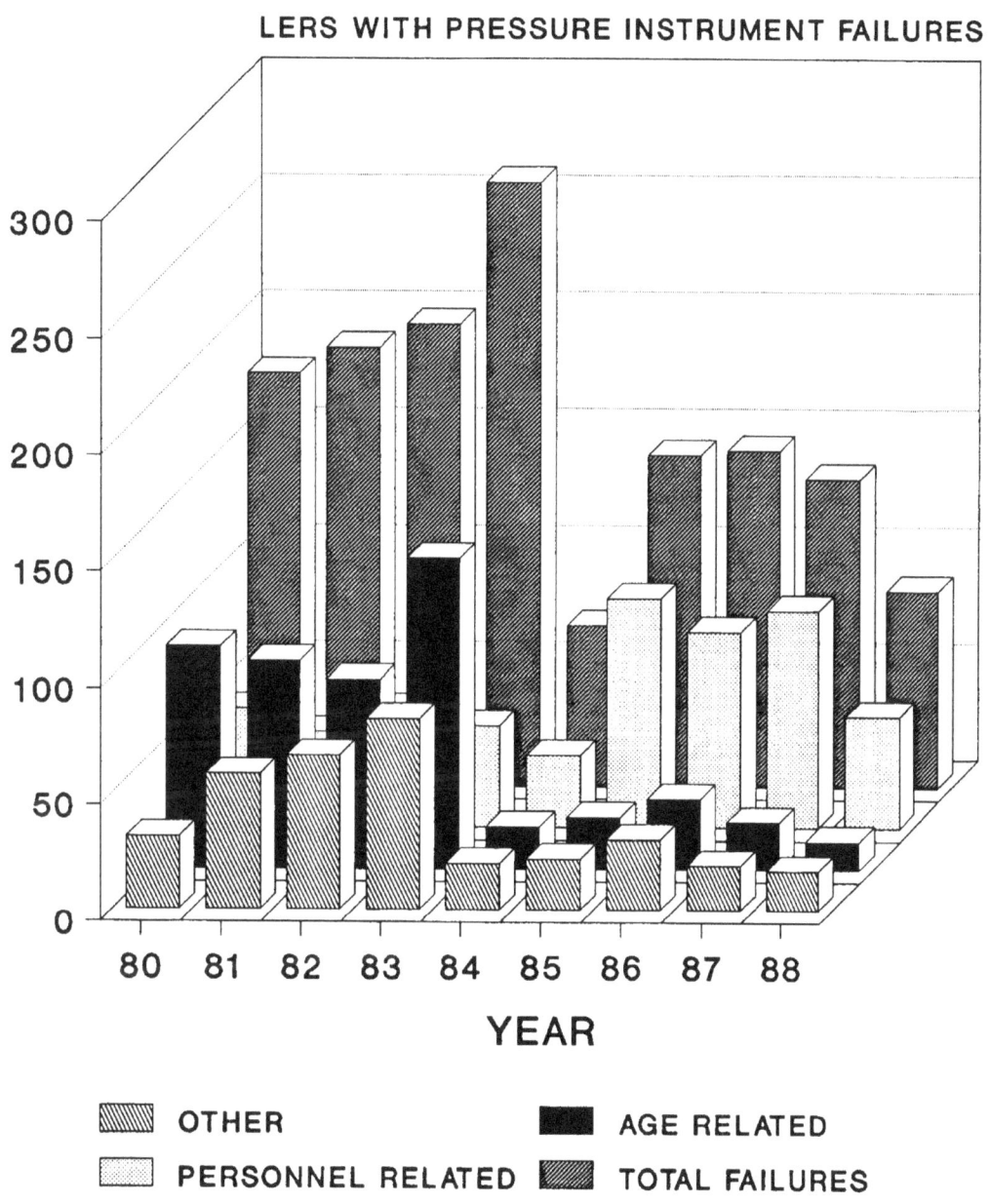

Figure 18.2.   Illustration of LERs Reporting
Pressure Transmitter Problems.

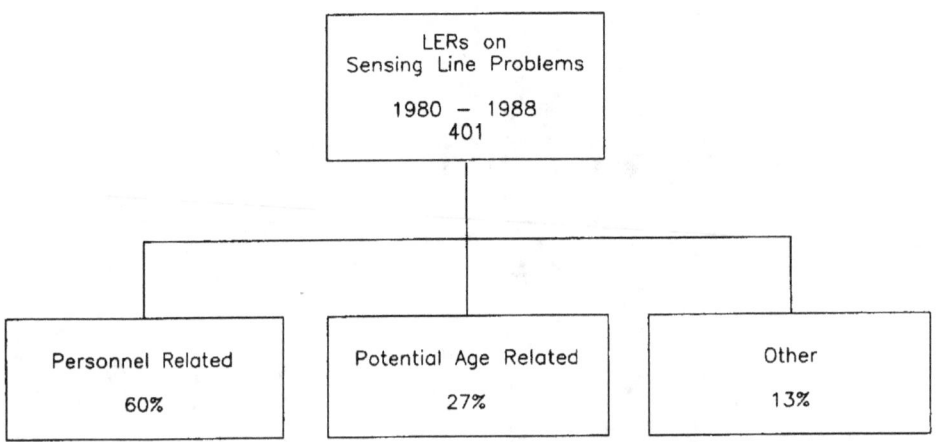

Figure 18.3. Breakdown of Sensing Line LERs.

TABLE 18.1

SAMPLE LER ABSTRACTS REPORTING
TRANSMITTER RESPONSE TIME PROBLEMS

****

DOCKET	YEAR	LER NUMBER	REVISION	PLANT NAME	UNIT	EVENT DATE
260	83	063	1	BROWNS FERRY	2	10/07/83

****

ON 10/13/83, DURING ROUTINE ANALYSIS OF THE 10/07/83 UNIT 2 SCRAM, IT WAS DISCOVERED THAT 2-LT-85-45A (WEST SIDE SCRAM DISCHARGE INSTRUMENT VOLUME (SDIV) LEVEL TRANSMITTER) DID NOT INITIATE A TRIP WHEN THE SDIV LEVEL EXCEEDED THE 50-GALLON SETPOINT (TECH SPEC TABLE 3.1.A). ON 10/14/83 TESTING REVEALED SLOW AND ERRATIC RESPONSE TO THE MEASURED VARIABLE. REDUNDANT LEVEL SWITCH (LS-85-45C) WAS OPERABLE. SCRAM CHANNEL 'A' WAS MANUALLY TRIPPED AT 1405 ON 10/13/83. THE ROSEMOUNT 1153DBN0005 TRANSMITTER (LT-85-45A) WAS REPLACED. SMALL PUNCTURE HOLES WERE FOUND IN THE TRANSMITTER'S MODULE DIAPHRAGM WHICH CAUSED THE SLOW RESPONSE TO MEASURED VARIABLE. THE REPLACEMENT TRANSMITTER WAS CALIBRATED, TESTED, AND PLACED IN SERVICE, PER SI 4.1.A-8. SCRAM CHANNEL 'A' WAS THEN RESET.

****

DOCKET	YEAR	LER NUMBER	REVISION	PLANT NAME	UNIT	EVENT DATE
325	83	058	0	BRUNSWICK	1	11/10/83

****

DURING UNIT SHUTDOWN OPERATIONS, PERFORMANCE OF REACTOR HIGH PRESSURE TRIP RESPONSE TIME TEST, PT-A26.2, REVEALED THAT REACTOR VESSEL STEAM DOME PRESSURE HIGH INSTRUMENT, B21-PT-N023D, ACTUATED OUT OF SPECIFICATIONS (0.59 SECONDS VERSUS THE TECH SPEC REQUIREMENT OF LESS THAN OR EQUAL TO 0.55 SECONDS), DUE TO AN INCORRECT INSTRUMENT DAMPING ADJUSTMENT. THIS INSTRUMENT, WHICH IS RESPONSE TIME TESTED EVERY 72 MONTHS, WAS PLACED IN SERVICE ON JUN 27, 1981. TECH SPECS 3.3.1, 6.9.1.9A. THE INCORRECT INSTRUMENT DAMPING ADJUSTMENT IS BELIEVED TO HAVE BEEN MADE BY THE INSTRUMENT MANUFACTURER. THE DAMPING ADJUSTMENT SETTING WAS READJUSTED AND THE N023D, MODEL NO. 1152, TESTED SATISFACTORILY. THE RESPECTIVE INSTRUMENT DAMPING ADJUSTMENTS OF 75 UNIT NO. 1, 52 UNIT NO. 2, AND 44 IN PLANT STOCK (MODEL NO. 1152) INSTRUMENTS WERE CHECKED. ONE OF THESE INSTRUMENTS, 2-B21-LT-N042D, WAS FOUND TO HAVE IMPROPERLY POSITIONED DAMPENING ADJUSTMENTS.

---

continued on next page

TABLE 18.1 (continued)

```
********************************************************************************
```
DOCKET	YEAR	LER NUMBER	REVISION	PLANT NAME	UNIT	EVENT DATE
306	86	002	0	PRAIRIE ISLAND	2	05/20/86
```
********************************************************************************
```

POWER LEVEL - 019%. ON MAY 20, 1986, UNIT 2 WAS BEING RESTARTED FROM THE TRIP OF MAY 19, 1986. THE UNIT WAS SYNCHRONIZED WITH THE GRID AT 0258 AND FEEDWATER CONTROL WAS TRANSFERRED FROM THE BYPASS VALVES TO THE MAIN REGULATING VALVES. THE CONTROLLERS FOR BOTH MAIN FEEDWATER REGULATING VALVES WERE PLACED IN AUTOMATIC WITHOUT INCIDENT AND WERE CLOSELY MONITORED BECAUSE OF PREVIOUS INSTANCES OF ERRATIC OPERATION AT LOW POWER LEVELS. IT WAS NOTED THAT NO. 12 STEAM GENERATOR WAS BEING OVERFED, AND THE FEEDWATER CONTROL SYSTEM WAS TRANSFERRED BACK TO MANUAL, BUT AT 0319 THE UNIT TRIPPED ON HIGH STEAM GENERATOR LEVEL. PRIMARY CAUSE OF THE EVENT IS FAILURE OF THE FEEDWATER CONTROL SYSTEM TO ADEQUATELY CONTROL STEAM GENERATOR LEVEL AT LOW POWER LEVEL. A CONTRIBUTING CAUSE TO THIS EVENT WAS THE SLUGGISH RESPONSE OF STEAM GENERATOR LEVEL TRANSMITTERS. HARDWARE HAS BEEN PURCHASED THAT WILL PROVIDE BETTER FEEDWATER FLOW INDICATION AT LOW POWER. INSTALLATION IS PLANNED AT THE NEXT REFUELING. THE STEAM GENERATOR LEVEL TRANSMITTER VARIABLE AND REFERENCE LEGS WILL BE BLOWN DOWN MORE FREQUENTLY. DIFFICULTY CONTROLLING STEAM GENERATOR LEVEL AT LOW POWER LEVEL HAS RESULTED IN SEVERAL REACTOR TRIPS ON BOTH UNITS. SIMILAR EVENTS ARE DESCRIBED IN UNIT 1 LER'S 86-005, 84-011, 84-006 AND 84-001.

```
********************************************************************************
```
DOCKET	YEAR	LER NUMBER	REVISION	PLANT NAME	UNIT	EVENT DATE
306	83	022	0	PRAIRIE ISLAND	2	09/09/83
```
********************************************************************************
```

DURING SURVEILLANCE TEST, ONE STEAM FLOW TRANSMITTER WAS FOUND OUT OF CALIBRATION LOW BY 1%. TRANSMITTER RESPONSE WAS ALSO SLUGGISH. REDUNDANT EQUIPMENT WAS OPERABLE. TECH SPEC TABLES 3.5-2 AND 3.5-4 APPLY. CAUSE NOT KNOWN AT THIS TIME; FAILED TRANSMITTER WILL BE RETURNED TO THE VENDOR FOR ANALYSIS. THE ROSEMOUNT MODEL 1153HA6 DIFFERENTIAL PRESSURE TRANSMITTER HAS BEEN REPLACED.

TABLE 18.2

SAMPLE LER ABSTRACTS REPORTING
SENSING LINE PROBLEMS

****

DOCKET	YEAR	LER NUMBER	REVISION	PLANT NAME	UNIT	EVENT DATE
244	88	002	0	GINNA		03/08/88

****

POWER LEVEL - 000%. ON MARCH 8, 1988 AT 0135 EST WITH THE REACTOR COOLANT SYSTEM TEMPERATURE AND PRESSURE AT 340F AND 350 PSIG RESPECTIVELY, THE CONTROL ROOM OPERATORS OBSERVED THAT THE LEVEL OF THE ONE OPERABLE BORIC ACID STORAGE TANK WAS INDICATING LESS THAN THE TECH SPEC REQUIREMENT FOR ONE OPERABLE BORIC ACID STORAGE LEVEL TANK AND CONSTITUTED A PLANT CONDITION PROHIBITED BY TECH SPECS. THE CONTROL ROOM OPERATORS TOOK IMMEDIATE CORRECTIVE ACTION AND RESTORED THE OPERABLE BORIC ACID STORAGE TANK LEVEL TO MEET THE TECH SPEC REQUIREMENTS WITHIN APPROXIMATELY 30 MINUTES. THE IMMEDIATE CAUSE OF THE EVENT WAS LEVEL INDICATION INACCURACIES DUE TO PARTIAL PLUGGING OF THE LEVEL SENSING LINES. THE CAUSE OF THIS PARTIAL PLUGGING IS POSSIBLY DUE TO THREE ROOT CAUSES. (1) NOT CLEANING THE SENSING LINES PRIOR TO DECLARING THE TANK OPERABLE, (2) TYPE OF LEVEL INSTRUMENTATION AND, (3) TOO LOW A TEMPERATURE NITROGEN GAS. CORRECTIVE ACTION TAKEN OR PLANNED TO PREVENT RECURRENCE IS TO REQUIRE CLEANING SENSING LINES PRIOR TO DECLARING TANK OPERABLE AND TO REVIEW THE APPLICATION OF THE EXISTING LEVEL INSTRUMENTATION AND THE NITROGEN GAS TEMPERATURE AND MAKE RECOMMENDATIONS AS NECESSARY.

---

continued on next page

TABLE 18.2 (continued)

```
****************************************************************************
```

DOCKET	YEAR	LER NUMBER	REVISION	PLANT NAME	UNIT	EVENT DATE
325	83	063	1	BRUNSWICK	1	12/03/83

```
****************************************************************************
```

SURVEILLANCE DURING UNIT POWER OPERATION REVEALED REACTOR WATER CLEANUP SYSTEM (RWCS) DIFFERENTIAL FLOW INDICATOR, 1-G31-R615, WAS SHOWING AN ERRONEOUS INDICATION OF RWCS DIFFERENTIAL FLOW. DURING SUBSEQUENT UNIT POWER OPERATION ON DECEMBER 7, 1983, SPURIOUS RWCS "LEAK HI-HI" ALARM ANNUNCIATIONS OCCURRED. THE EVENTS OCCURRED DUE TO ENTRAPPED AIR IN THE SENSING LINES OF THE RWCS LEAK DETECTION SYSTEM. THIS RESULTED FROM A PROCEDURAL INADEQUACY IN THE RWCS HIGH FLOW RESPONSE TIME TEST, PT-45.2.16, WHICH WAS PERFORMED RESPECTIVELY ON DECEMBER 3 AND 6, 1983. THE ENTRAPPED AIR WAS REMOVED AND THE RWCS LEAK DETECTION SYSTEM WAS RETURNED TO SERVICE. APPROPRIATE REVISIONS TO PT-45.2.16 WERE IMPLEMENTED TO HELP PREVENT FUTURE SIMILAR OCCURRENCES.

```
****************************************************************************
```

DOCKET	YEAR	LER NUMBER	REVISION	PLANT NAME	UNIT	EVENT DATE
261	85	015	0	ROBINSON	2	07/05/85

```
****************************************************************************
```

POWER LEVEL - 012%. A REACTOR TRIP OCCURRED ON 7-5-85, AT 1135 HRS. THE PLANT WAS AT 12% POWER REDUCING LOAD TO REPAIR 'A' FEEDWATER REGULATING VALVE. A STEAM FLOW GREATER THAN FEED FLOW (SF > FF) SIGNAL WAS GENERATED BY A STEAM FLOW SPIKE CAUSED BY REMOVING THE MOISTURE SEPARATOR REHEATERS FROM SERVICE AND A MALFUNCTION IN A MSR SHUTOFF VALVE. CONCURRENTLY, A 'C' SG LOW LEVEL SIGNAL WAS RECEIVED DUE TO PARTIAL BLOCKAGE OF THE REFERENCE LEG FOR SG LEVEL TRANSMITTER 494 AND THE SG PRESSURE DECREASE WHICH OCCURRED WHEN STEAM FLOW INCREASED. THE LOW SG LEVEL SIGNAL COINCIDENT WITH THE SF > FF SIGNAL RESULTED IN THE REACTOR TRIP. THE CONTROL VOLTAGE TRANSFORMER FOR THE 1B MSR SHUTOFF VALVE WAS REPLACED. LT-494 WAS REPAIRED BY BLOWING DOWN ITS SENSING LINES. A REV TO GP-006, 'NORMAL PLANT SHUTDOWN FROM POWER OPERATION TO HOT SHUTDOWN,' TO IMPROVE THE STEPS INVOLVED WITH REMOVING THE MSR'S FROM SERVICE WILL BE MADE BY 10-4-85. THE PLANT WAS RETURNED ON LINE AT 0345 HRS ON 7-6-85.

19. REVIEW OF REGULATORY GUIDES AND STANDARDS

The latest draft of the documents which relate to sensor response time testing in nuclear power plants are:

1. Regulatory Guide 1.118 entitled, "Periodic Testing of Electric Power and Protection Systems", Rev. 2, published by the U. S. Nuclear Regulatory Commission in 1978.

2. IEEE Standard 338 - 1987 entitled, "Standard Criteria for the Periodic Surveillance Testing of Nuclear Power Generating Station Safety Systems". This is a revision of IEEE Standard 338 - 1977 published by the Institute of Electrical and Electronics Engineers.

3. ISA Standard S67.06 entitled, "Response Time Testing of Nuclear Safety-Related Instrument Channels in Nuclear Power Plants", issued in final form in 1984 by the Instrument Society of America.

The original drafts of these documents were written about ten years ago, and there have been no new revisions except for the IEEE Standard which was revised in 1987. However, in relation with sensor response time measurements in nuclear power plants, the new version of the IEEE Standard is not significantly different than the 1977 version.

The ISA Standard S67.06 which was issued in 1984, was actually written several years earlier and is based predominantly on information that was available in the late 1970's and early 1980's. Furthermore, although this standard contains several specific criteria, a few sections of it are written in a manner than can be subject to a wide variety of interpretations.

The standards mentioned above and the Regulatory Guide 1.118 do not reflect the state-of-the-art in sensor performance testing, nor do they address new problems and concerns such as sensing line clogging problems and aging concerns which have come to light since these documents were prepared. An example of a specific and important problem in all three documents is related to pressure sensing lines in nuclear power plants. The IEEE Standard 338 states, in Section 6.3.4 paragraph (6), that response time testing of the process to sensor coupling is not required. This is in direct conflict with ISA Standard 67.06 which specifies in Section 5 that the sensing lines should be tested. With regard to this point, Regulatory Guide 1.118 questions the IEEE Standard and specifies in Section C.8 of the Regulatory Guide that the NRC will study the IEEE provision and will provide the results in a future draft of Regulatory Guide 1.118. However, a new draft of the Regulatory Guide 1.118 has not been written to clarify the NRC's concern about sensing line delays.

The IEEE and ISA Standards and the regulatory guide mentioned above need to be revised in light of new testing technologies which are now available. For example, the power interrupt method has been developed and validated for on-line response time testing of force-balance pressure transmitters. Although this method has been recognized since the early 1980's, there is no mention of it in the ISA Standard. Also lacking in the ISA Standard is an effective discussion of test uncertainties, and there is also no mention of a need to document the data reduction methods and procedures.

For acceptance criteria, there are general statements in these documents that the response time results must satisfy technical specification requirements. However, many plants and especially older plants, do not have clear provisions in their technical specifications as how to address sensor response time. Sensor response time testing technologies did not exist when the original draft of technical specifications of many plants were written. The standards or the Regulatory Guide should address the adequacy of current technical specifications and provide guidelines with respect to response time testing of safety system sensors. This is important because a sensor is a vulnerable component of an instrument channel which is susceptible to aging degradation, especially since it is usually located in a harsh environment.

20. REVIEW OF RELATED RESEARCH

There has been a number of studies on aging effects on performance of nuclear plant pressure transmitters, most of which concentrated on static performance (i.e., calibration drift). Sandia National Laboratories has performed several aging studies on nuclear safety-related equipment. This includes an experimental study with five Barton Model 763 pressure transmitters[7]. These transmitters were tested to determine the failure and degradation modes in separate and simultaneous environmental exposures. This study shows that temperature is the primary environmental stressor affecting the static performance of the Barton transmitters tested. Also performed at Sandia was work on several Barton and Foxboro pressure transmitters which were removed from Beznau Nuclear Power Station in Switzerland and sent to Sandia for testing[8]. These transmitters had aged naturally in the plant for eight to twelve years. This experimental work showed that some degradation had occurred, but concluded that the performance of the transmitters remained satisfactorily.

Idaho National Engineering Laboratory has also reviewed the performance of nuclear plant pressure transmitters. In a report on the evaluation of operating experiences of nuclear power plants published in January 1988, Leroy Meyer of EG&G concluded (from a review of Nuclear Power Plant Experience (NPE) data base for Reactor Trip Systems) that components associated with pressure measurements experience the highest number of failure events[9]. An important point in this report (which relates to this project) is that pressure transmitter response time testing should be performed in Light Water Reactors (LWRs) and that the response time of the sensing lines should be included. It also states that when adequate research is completed to determine optimum test frequencies, Regulatory Guide 1.118 should be revised. This work was done as a part of the Nuclear Plant Aging Research (NPAR) program sponsored by the NRC[10]. Also performed under NPAR program is a study by Gary Toman of Franklin Research Center (FRC), which involved an evaluation of the stresses that cause degradation in nuclear plant pressure transmitters[11]. The report on this work also describes a means of detecting and evaluating the degradation of pressure transmitters. Gary Toman's study concluded that the major consequences of the stresses on pressure transmitters are calibration shifts. Both the EG&G and the FRC's work were predominantly review studies.

A few studies have been done on dynamic performance of pressure transmitters. This includes work at Oak Ridge National Laboratory (ORNL) by Mullens and Thie, who performed a general study of nuclear reactor pressure noise. This work involved theoretical and experimental research, as well as a literature review, to determine the potential of pressure fluctuations for on-line monitoring of reactor behavior including the response time behavior of pressure transmitters and the associated sensing lines[12]. In this study, which was published in 1985, Mullens and Thie conducted a search of the LER data base and found that sensing lines have contributed significantly to dynamic failures of pressure sensing systems in nuclear power plants. In other experimental research at ORNL, Mike Buchanan and others performed an evaluation of the methods for measurement of response time and detection of degradation in pressure sensor and sensing line systems[4]. The methods studied were noise analysis and power interrupt (PI) tests, which were discussed earlier in this report. The ORNL team concluded that both the noise analysis and the PI tests are suitable for response time testing of pressure transmitters in nuclear power plants.

Tennessee Valley Authority (TVA) has also performed studies on pressure sensor dynamics. In a paper published in November 1987, Gerald Schohl of TVA has reported the results of an experimental study on the effects of sensing line air on pressure measurements in nuclear power plants. He has shown that air in sensing lines can affect pressure sensing system response time[13]. Work on sensing line problems has also been performed at EG&G Idaho. In a paper by R. P. Evans and G. G. Neff, EG&G reported that the effects caused by unequal sensing line lengths in differential pressure transmitters can cause error in pressure measurements and (during rapid pressure transients) could cause the transmitter to fail[14]. This paper reported on experimental work involving the Loss-of-Fluid Test (LOFT) facility. The same authors have written another paper entitled "Line Pressure Effects on Differential Pressure Measurements". This work reports on another experimental study which was concerned with static pressure effects on sensor calibration[15].

Work on pressure sensors sponsored by others includes recent studies on the problem called "oil leak" syndrome in Rosemount transmitters, which is described in an NRC information notice given in Appendix B. This problem is being investigated by two utilities (Northeast Utilities and Public Service Electric and Gas Company of New Jersey), the transmitter manufacturer, and AMS. These studies have involved laboratory and field measurements with Rosemount Model 1153 and 1154 transmitters.

21. CONCLUSIONS

An experimental assessment of the methods available for response time testing of nuclear plant pressure transmitters was successfully performed. The assessment concluded that the five methods that are available are equally effective if used properly. The normal accuracy of these methods is estimated to be about 100 milliseconds. This accounts for the uncertainties in performing the tests and the uncertainties associated with the repeatability characteristics of pressure transmitters. The five test methods are referred to as ramp test, step test, frequency test, noise analysis and power interrupt test. Two of these methods (noise analysis and power interrupt test) have the advantage of providing on-line measurement capability at normal operating conditions.

A laboratory study of normal aging effects on performance of nuclear-type pressure transmitters was initiated in this project and preliminary results were obtained. The study involved exposure of several pressure transmitters to heat, humidity, vibration, pressure cycling, and overpressurization conditions. The results showed calibration shifts and response time degradations with the former being the more pronounced problem. Thermal aging was found as one of the more important causes of performance degradation in pressure transmitters, with the transmitter's electronics being more susceptible to thermal aging than its other components.

A search of the NRC's Sequence Coding and Search System LER database for pressure sensing system problems was performed. The search revealed 1,325 cases of reported problems with pressure sensing systems over a nine-year period beginning with 1980. Potential age-related cases accounted for 38 percent of the reported problems in this period. A notable number of the LERs were related to sensing line problems such as blockages, freezing, and void in the line.

Another aspect of this project was a preliminary review of the Regulatory Guide 1.118, IEEE Standard 338, and ISA Standard 67.06. The review indicated that both the Regulatory Guide and the two Standards should be revised to account for recent advances in performance testing technologies and other information that have become available since these documents were written about ten years ago.

REFERENCES

1. U.S. Nuclear Regulatory Commission, "Periodic Testing of Electric Power and Protection Systems," Regulatory Guide 1.118, Rev. 2 (June 1978).

2. Foster, C. G., et.al., "Sensor Response Time Verification," Report No. NP-267, Electric Power Research Institute, Palo Alto, California (October 1976).

3. Currie, R.L., Mayo, C. W., Stevens, D.M., "ARMA Sensor Response Time Analysis," Electric Power Research Institute, Report Number NP-1166 (May 1978).

4. Buchanan, M.E., et.al., "Measurement of Response Time and Detection of Degradation in Pressure Sensor/Sensing Line Systems," NUREG/CR-4256, Oak Ridge National Laboratory Report Number ORNL/TM-9574 (September 1985).

5. Rosemount Inc., "Nuclear Equipment Analysis Report, Rosemount Model 1153 Service B Pressure Transmitters," Rosemount Report 57820, Rev. D, 1978.

6. Keenan, M. R., "Moisture Permeation Into Nuclear Reactor Pressure Transmitters," Sandia National Laboratories, Report No. SAND-83-2165.

7. Furgal, D.T., Craft, C.M., Salzar, E.A., "Assessment of Class 1E Pressure Transmitter Response When Subjected to Harsh Environment Screening Tests," NUREG/CR-3863, Sandia National Laboratories, Report Number SAND 84-1264 (March 1985).

8. Grossman, J.W., Gilmore, T.W., "Evaluation of Ambient Aged Electronic Transmitters from Beznau Nuclear Power Station," NUREG/CR-4854, Sandia National Laboratories Report Number SAND 86-2961, Draft, May 1988, available in NRC Public Document Room, 2120 L Street, NW, Washington, DC 20555.

9. Meyer, L.C., "Nuclear Plant-Aging Research on Reactor Protection Systems," NUREG/CR-4740, EG&G Idaho, Inc., Report EGG-2467 (January 1988).

10. U.S. Nuclear Regulatory Commission, "Nuclear Plant Aging Research (NPAR) Program Plan," NUREG-1144 (September 1987).

11. Toman, G.J., "Inspection, Surveillance, and Monitoring of Electrical Equipment in Nuclear Power Plants, Volume 2. Pressure Transmitters," NUREG/CR-4257 VOl. 2, Oak Ridge National Laboratory Report Number ORNL/Sub/83-28915/3/V2. (August 1986).

12. Mullens, J.A., Thie, J.A., "Pressure Noise in Pressurized Water Reactors," Oak Ridge National Laboratory Report No. ORNL/TM-9773, NUREG/CR-4389 (December 1985).

13. Schohl, G.A., et.al., "Detection of Air in Sensing Lines from Standing Wave Frequencies," Transactions of American Nuclear Society, 1987 Annual Winter Meeting, Los Angeles, California (November 1987).

14. Evans, R.P., Neff, G.G., "Sensing Line Effects on PWR-Based Differential Pressure Measurements," EG&G Idaho, Inc., EGG.M-01182 Preprint (February 1982).

15. Neff, G.G., Evans, R.P., "Line Pressure Effects on Differential Pressure Measurements," EG&G Idaho, Inc., EGG.M-02482 Preprint, ISA (1982).

APPENDIX A

AGE RELATED FAILURES OF PRESSURE
SENSING SYSTEMS IN NUCLEAR POWER PLANTS

ANALYSIS AND
MEASUREMENT SERVICES
CORPORATION

9111 CROSS PARK DRIVE NW / KNOXVILLE, TN 37923-4599 / (615) 691-1756

Report #: <u>NRC8901R1</u>

Age Related Failures of
Pressure Sensing Systems
in Nuclear Power Plants

Revision 1
May 1989

Prepared by

Analysis and Measurement Services Corporation
AMS 9111 Cross Park Drive, NW
Knoxville, Tennessee 37923-4599

Prepared for

U.S. Nuclear Regulatory Commission
Office of Nuclear Regulatory Research
Division of Engineering Technology

SUMMARY

A review of 1325 LERs reporting failures of pressure instrumentation over the 1980 through October 1988 time period was made. The review identified 498 of these 1325 LERs as having age-related failures.

The predominate age-related effects were setpoint drift or calibration problems (reported in about 45% of the LERs); water spray, condensation, flow blockages, or freezing (reported in about 38% of the LERs); and worn, bent, broken, or damaged subcomponents (reported in about 15% of the LERs). Corrosion, erosion, vibration, and fatigue made up the remaining few percent.

The review also identified 540 LERs where the pressure instrumentation problems were due to personnel errors and 342 LERs which occurred for other, or unknown reasons.

Component vendor codes associated with age-related failures were examined. Barton, Fischer, Rosemount, and Foxboro were the most frequently occurring vendor codes (about 10 to 15% each). No vendor code was given in about 20% of the cases and about 16% of the failures were attributed to an "other" category consisting of about 20 vendors.

Westinghouse and Babcock and Wilcox NSSS plants reported a higher number of pressure, level, or flow instrumentation problems per plant than did Combustion Engineering or General Electric plants.

Sensing line problems contributed to about 400 LERs reporting pressure instrumentation problems, with about 60% due to personnel actions. Age-related problems (mostly freezing, condensation, or crud buildup) contributed to about 27% of the LERs.

TABLE OF CONTENTS

LIST OF TABLES

LIST OF FIGURES

1. INTRODUCTION

This report presents the results of a review of licensee event reports (LERs) which reported problems with pressure, level, or flow instrumentation, with particular emphasis given to age-related failures. Age-related failures are defined to be those failures attributable to effects of time. Time effects are particularly important when components are operated in environments where cyclic pressures or temperatures, humidity, vibration, or corrosive or erosive conditions and many others may serve to degrade component performance.

Pressure, level, and flow instruments are defined as those components required to measure or sense pertinent changes in process system variables. These components include sensing lines and associated root or isolation valves, sensor primary elements, transmitters, switches, cables, amplifiers and other signal conditioning components, controllers, and indicators. The collection of these components required to communicate process conditions will be referred to hereafter as a pressure, level, or flow instrumentation systems, or pressure instruments for short, since pressure, level, and flow instrumentation frequently depend upon pressure signals initially and use many of the same signal conditioning components.

LERs reporting pressure instrumentation problems were identified and categorized using the Sequence Coding and Search System (SCSS) LER data base operated and maintained by the Nuclear Operations Analysis Center (NOAC) at the Oak Ridge National Laboratory (ORNL). As of mid-January, the SCSS data base contained LERs with event dates from 1980 approximately through October 1988.
The SCSS database was used to identify events where (1) pressure, level, or flow instrumentation problems required repair replacement, calibration, or other action to restore function, (2) sensing line problems affected pressure, level, or flow instrumentation, or (3) isolation or root valve problems affected pressure, level, or flow instrumentation. Section 2 of this report describes the development of the SCSS searches.

Section 3 discusses the applicability of LER information to pressure instrumentation problems. The effects of the change in LER reporting requirements which took place in 1984 are also discussed.

Section 4 presents the results of the review. First, the numbers of LERs reporting pressure instrumentation problems related to personnel, age, or other factors are presented and briefly discussed. Additional characterizations for pressure instrumentation problems in (1) the reactor protection system (RPS), engineered safety features (ESF) actuation system, or primary coolant system leakage detection systems and (2) other systems were made in order to determine whether problems affecting pressure instrumentation were system dependent.

Problems affecting the instrumentation were then characterized by the type of condition contributing the problems: personnel errors, age-related effects, or unknown (or other) problems. The most frequently occurring types of potential age-related effects were examined. Age-related failures of pressure instruments were also categorized by NSSS vendor to identify whether some vendors appear to be more susceptible to these failures than others. A list of component vendors was compiled from LERs which reported pressure instrument failures to complete the descriptive information.

Pressure instrumentation problems caused by sensing line effects were examined next. The examination included a breakdown by personnel related, age-related, or unknown (or other) cause factors. The problems were further characterized to determine the most frequently occurring age-related effects.

Instrumentation problems resulting from instrument root or isolation valves failures or improper configurations were also specifically reviewed for errors related to personnel actions, age, or unknown (or other) reasons.

2. REVIEW METHODOLOGY

2.1 Description of SCSS Data Base

The Sequence Coding and Search System (SCSS) data base of LER information formed the basis for the categorization of pressure instrumentation failure data. This data base was developed by the NRC Office for Analysis and Evaluation of Operational Data (AEOD). The SCSS data base is operated for AEOD by the Nuclear Operations Analysis Center (NOAC) of the Oak Ridge National Laboratory (ORNL). The SCSS data base contained data from LERs with event dates from 1980 approximately through the end of October 1988 (about 30,000 LERs) as of the middle of January 1989.

The SCSS data base was developed in order to allow information reported in the LER and accompanying descriptive text to be encoded such that detailed, comprehensive information regarding component and system failures, personnel errors, and unit effects and their interactions could be retrieved. The structure of the data base permits the identification of events where personnel actions affect certain systems or components. These, in turn, may have their own effects on other components or systems. Ultimately, they may even cause reactor trips or ESF actuations. Several features of the data base which have particular relevance to this subject include:

(1)   The function of the instrumentation system (pressure, flow, level, radiation, etc.) and the type of component (primary elements, indicators, controllers, etc.) are encoded when specified.

(2)   The cause of component failures are specified using codes amenable to identifying potential age-related failures.

(3)   The particular instrument system (RPS, feedwater control, etc.) is specified.

(4)   The timing of component failures is specified (i.e., pre-existing failure, immediate failure, etc.).

(5)   The need for component repair action is stated.

(6)   The component vendor is specified when so stated by the licensee in the LER.

(7)   Relationships between the sensor failures and prior component failures or personnel errors are noted.

These features of the data base made it's use attractive to evaluate pressure instrumentation failures. The intent of the review was to focus only on the sensor or sensing lines. However, due to different levels of specificity used by utilities in preparing LER information, the review was kept at relatively broad instrumentation functional areas (flow, level, etc.) rather than initially focus only on sensors or sensing lines.

2.2 Description of SCSS Data Base Searches

Search strategies covering three areas were developed to encompass the types of instrument appropriate for this review. These strategies covered the following areas:

A - 2

(1)   Pressure, level, and flow element failures which required some form of repair or corrective action (replace, adjust, clean) to fix.

(2)   Pressure, level, and flow element failures which linked to previous problems with sensing or impulse line.

(3)   Pressure, level, and flow element failures which linked to previous problems with instrument root or isolation valves.

The LERs which were retrieved using search strategies for the three categories above were then evaluated for effects related to personnel, age, or some other factor. These evaluations were made using selections of cause, effect, and linkage codes available through SCSS data base commands.

2.2.1 Age-related failures. Potential age-related problems affecting pressure instrumentation were retrieved based upon the occurrence of cause or effect codes covering the following subject areas:

-     Worn, bent, damaged, or broken components;
-     Vibration, mechanical fatigue, or thermal fatigue;
-     Boron precipitation, freezing conditions, flow blockages, or water spray or condensation;
-     Drift or calibration problems;
-     Corrosion or erosion.

LERs exhibiting pressure instrumentation problems related to any of the above categories were noted as potentially age-related.

2.2.2 Personnel related. LERs where the pressure instrumentation problems were connected to personnel errors were noted as personnel related. It is possible for LERs to contain both age-related and personnel related causes. LERs reporting both potential age-related and personnel related failures were ultimately categorized as personnel errors.

2.2.3 Other. LERs which reported pressure instrumentation failures where there were no connections to age or personnel related causes were categorized as "Other", which includes unknown or unstated causes.

3. APPLICABILITY OF LER DATA

3.1 Change in LER Reporting Requirements in 1984

LERs have been required of the nuclear industry since the early 1970s to report certain types of operational problems affecting commercial nuclear power plants. From that time through 1983, the information reported in LERs and the format in which it was transmitted to the NRC remained relatively constant. The reportability requirements during this time depended on federal regulations and plant technical specifications or provisions in their license.

The NRC issued revised reporting criteria effective January 1, 1984. This had a significant effect on LER information. This rule resulted in an approximately 50% decrease in the number of LERs reported in 1984 from those reported in 1983. The intent of the revised LER reporting requirements was to have more comprehensive information provided on significant types of events and reduce the requirements to report less significant events. More information was reported on ESF actuations and certain other types of events. LERs were no longer required for problems affecting only single components or instruments. The NRC justified this action in part because of the voluntary reporting of component failures to the Nuclear Plant Reliability Data System (NPRDS) operated by the Institute of Nuclear Power Operations (INPO).

Single component failures, especially those in systems which provided RPS or ESF actuations, were frequently required to be reported prior to the revised LER reporting requirements in 1984 because they were technical specification violations. LERs were frequently required when even a single instrument was found to be out of calibration. These types of events were no longer reportable as LERs unless the problems existed in multiple safety system instrumentation channels or unless they rendered a whole system inoperable. Thus, fewer LERs were submitted solely to report problems with a small number of instruments.

3.2 Applicability of LER Data

Even though the change in LER reporting requirements significantly reduced the number of LERs reporting instrumentation problems, widespread or pervasive problems noted by plant staff, whose effects crossed multiple system boundaries, were still reportable. Also reportable were those failures which were identified during, or contributing to, an event which did meet the requirements for an LER. It is presumed that while the number of various types of causes of instrument failures is reduced, the types of instrument failures exhibited in reportable events since 1984 are representative of those experienced under the previous reporting requirements, although in varying numbers and proportions.

4. RESULTS OF EVALUATION

The searches of the SCSS LER database identified 1325 LERs over the 1980 through October 1988 time period which reported pressure instrumentation problems. Age-related instrumentation problems were reported in 498 LERs. Personnel related instrumentation problems were reported in 540 LERs. Instrumentation failures in the "Other", or unknown category were reported in 342 LERs. (The LERs add to more than 1325 because LERs may report multiple instrumentation problems. Each of the problems is treated independently; multiple categories may be present in some LERs.) At the time these searches were performed, there were about 30,000 LERs on the LER data base. A non-trivial fraction of these LERs (1,325/30,000 = ~4%) report pressure instrumentation problems. About 40% of these are potentially age-related. The numbers of events merit additional investigation.

The search results were further reviewed for evidence of unusual counts or trends based upon the type of system affected, the distribution of cause codes, the vendor of the components experiencing failures, and the NSSS vendors. The results of these additional breakdowns are discussed in the sections which follow.

4.1 General Sensor Failure Data

Figure 1 shows yearly totals of LERs reporting pressure instrumentation failures over the 1980 to October 1988 time periods. (The data for 1988 is normalized over the whole year to facilitate yearly comparisons.) The effect of the 1984 LER rule is readily apparent. An analysis of the pre-1984 data indicates an upward trend in LER numbers which is partially due to increased numbers of "Other", or unknown failures and an increased number of age-related failures in 1983. The number of personnel errors which contributed to the instrumentation problems remained nearly constant over this period. Age-related failures made up approximately 40 to 50 percent of the total during this period. Approximately 25% of these age-related failures were setpoint drift.

The 1984 and newer data illustrates several interesting points. The most obvious is the factor of 3 to 4 drop in the number of LERs reporting pressure instrumentation problems from 1983 to 1984. The pronounced increase from 1984 to 1985 may be an adjustment period, or learning period, by licensees as they began to learn and report to the new requirements.

Another observation is the sharp and sustained reduction in age-related instrumentation failures, about a factor of 4 drop. The fraction of setpoint drift causes within this reduced group also decreased to about 10%. This reduction is almost certainly due to the change in reporting requirements for 1984 data. That is, many single instrument failures were no longer required to be reported as LERs in 1984.

Another interesting observation about the 1984 and newer data is the increase in the number of instrumentation failures contributed to, at least in part, by personnel errors. Part of this increase may be due to more detailed and descriptive event descriptions provided by the licensee as a result of the change in reporting requirements. A higher fraction of the events reportable as LERs were more complex or significant. These more complex or significant events more frequently report personnel errors.

4.2 Failures of Instrumentation in Important Actuation Instrumentation Systems

Figure 2 and Figure 3 show yearly totals of pressure instrumentation failures. Figure 2 shows yearly totals for pressure instrumentation in RPS actuation, ESF actuation, or primary coolant leakage detection

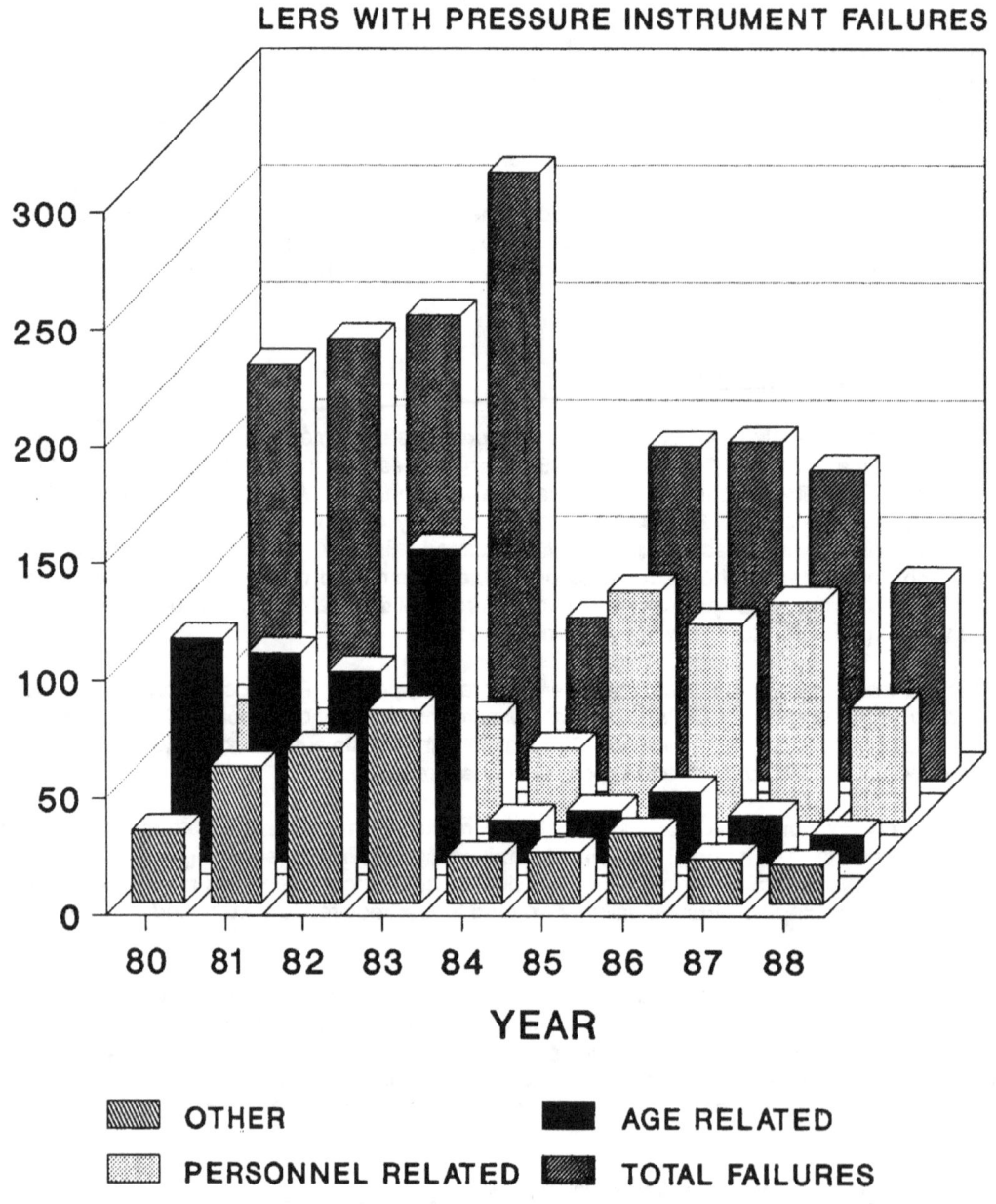

Figure 1

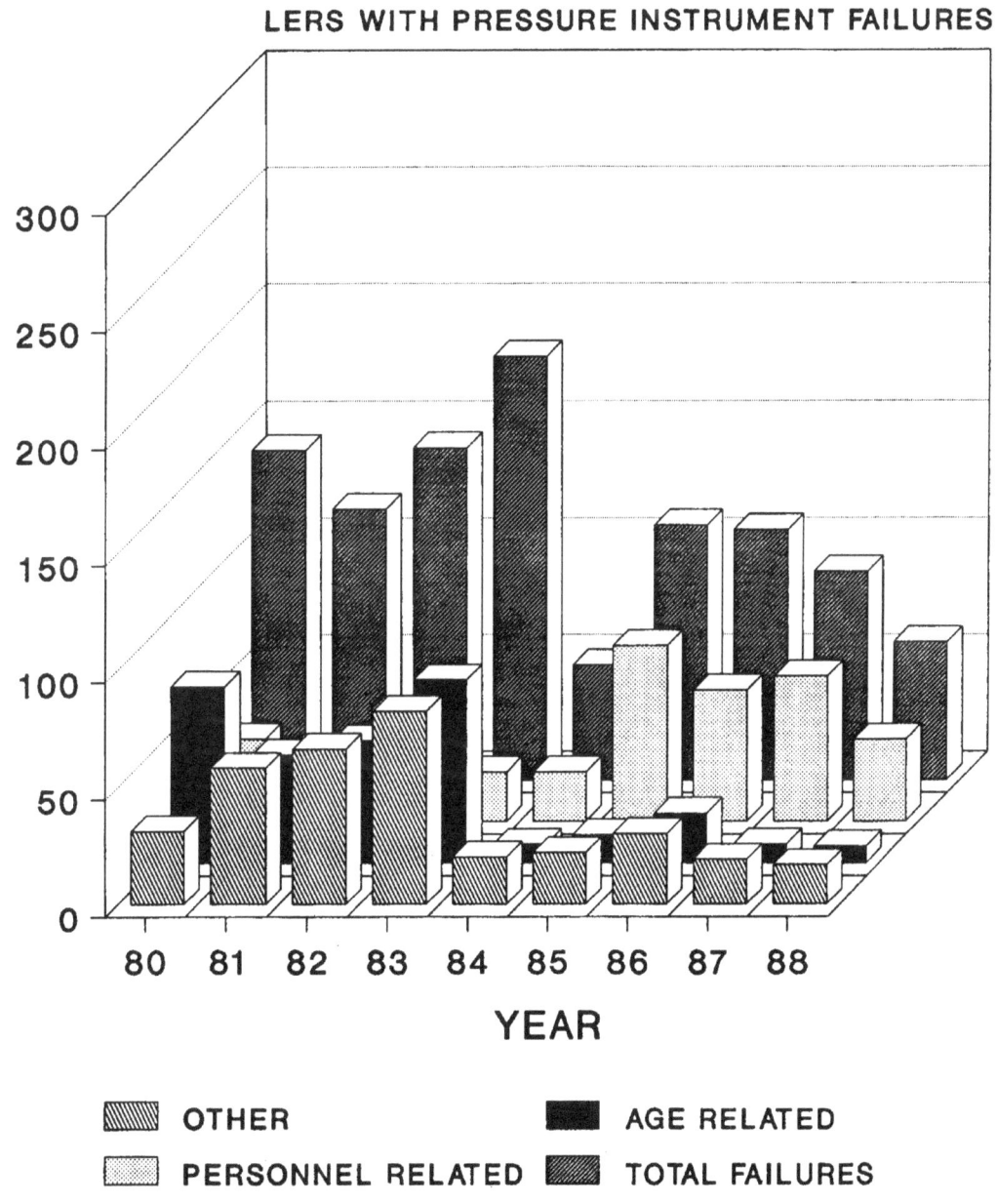

Figure 2

LERS WITH PRESSURE INSTRUMENT FAILURES

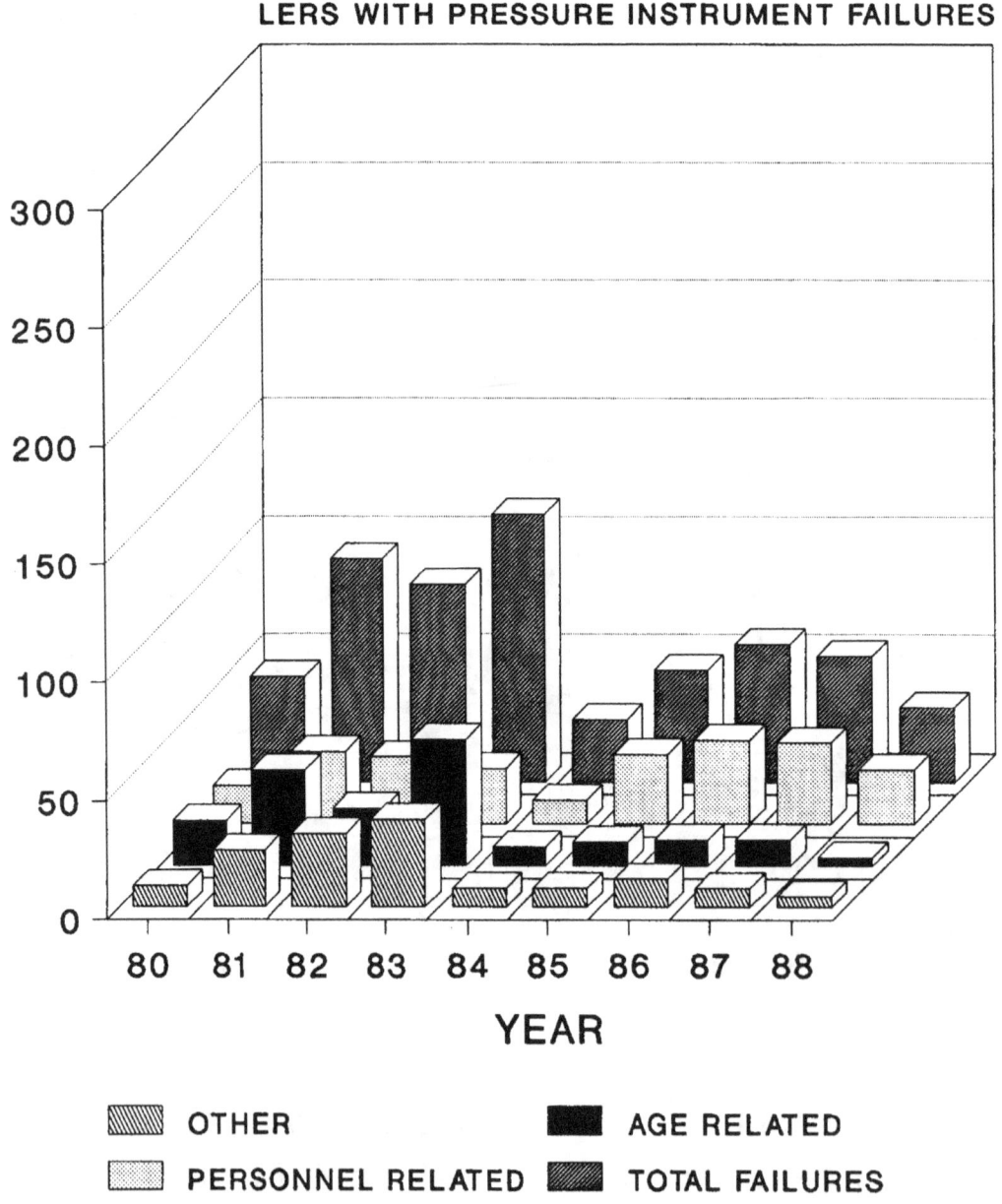

OTHER

PERSONNEL RELATED

AGE RELATED

TOTAL FAILURES

Figure 3

systems. Figure 3 shows yearly totals for pressure instrumentation in other systems. A brief examination of the two figures shows that instrumentation failures were reported about twice as frequently in the RPS actuation system, ESF actuation system, or primary coolant leakage detection system as in the other systems.

The most important observations to be made about Figures 1, 2, and 3 are that they all show similar behavior. That is, the figures show similar proportions of personnel, age, and other, or unknown, causes and that they show similar trends across the years. These observations are important because they indicate that pressure instrument problems are relatively independent of their system. This means that it will not be necessary to track multiple system breakdowns through the rest of the characterization.

4.3 Personnel Related Instrumentation Failures

Personnel errors contributed to 540 pressure instrumentation problems. The errors were not necessarily the root cause of the events, but a personnel error of some type preceded the instrumentation failure. Table 1 shows a breakdown of personnel activities which contributed to the personnel errors.

Design, maintenance, and testing or surveillance activities accounted for 70% of the personnel related LERs.

4.4 Age-Related Instrumentation Failures

Age-related failures of pressure instruments were identified based upon the occurrence of certain cause or effect codes used in the SCSS data base. One or more of these codes were

Table 1 Personnel Activities which Contributed to Instrument Failures

Personnel Activity	# of LERs	%
Administrative	20	4
Construction	9	2
Design	146	27
Fabrication	22	4
Installation	37	7
Maintenance	114	21
Operations	39	7
Testing/ Surveillance	122	22
Other/Unknown	27	6

present in 498 LERs. The categories for these codes were previously listed in Section 2.2.1. Table 2 shows the relative frequency of occurrence of the categories listed in Section 2.2.1. The codes used are not necessarily the root cause of the event, but they do indicate that conditions related to these codes contributed to the problems.

Problems related to setpoint drift and out of calibration accounted for about 45% of the age-related codes. The codes for water spray, condensation, freezing conditions, or flow blockages accounted for about 38%. The codes for worn, bent, broken, or damaged components accounted for about 14%. Other codes accounted for about 3%.

The age-related causes may effect instrument function in several areas. Obviously, an instrument that is out of calibration will be sending an erroneous signal to various controllers or indicators. Instruments suffering from almost any of the age-related categories may degrade instrument performance in this way. In addition, instrument performance may be degraded in a less visible manner. Each of the categories may result in degraded pressure instrumentation system response time to changes in process parameters. The time constant of individual components may understandably be affected by broken, damaged, or corroded mechanical components. Just as importantly; however, are degradations in the pressure instrumentation system. For example, a plugged or frozen sensing line would prevent pressure changes from even being apparent to pressure instrumentation. Thus, the response time to pressure perturbations could be exceeded even though the component response time of transmitters or signal conditioners are acceptable. Age-related effects must be considered as applying to the entire pressure instrumentation system as well as to the individual components.

The numbers shown in Table 2 are not specific LER counts, but relative frequencies of the cause categories. They take into account that a single LER may report more than one cause code. It is not uncommon to see more than one cause code used to describe pressure instrumentation failures within an LER.

The characterization of potential age-related pressure instrumentation failures is not meant to indicate exact percentages or numbers of failures, especially since the change in LER reporting requirements in 1984 reduced the number of reportable events. It does indicate, however, that age-related pressure instrument problems are not uncommon.

Table 2 Frequency of Age-Related Categories

Category	Frequency
Drift or calibration problem	45%
Water spray, condensation, flow blockage, or freezing	38
Worn, bent, broken, or damaged	14
Vibration or fatigue	1
Corrosion or erosion	2

4.5 Analysis by Component Vendor

LERs which reported age-related pressure instrumentation failures were examined for component vendor codes associated with these components. Figure 4 shows the vendor codes listed with these failures. Vendors were assigned to the failed instruments by the licensee in the LER.

As shown by the figure, vendor codes were specified by the licensee approximately 79% of the time. The most frequently occurring vendor codes were for Barton, Fischer, Rosemount, and Foxboro. Approximately 20 vendors accounted for the 16% shown as "Other."

4.6 Evaluation of Age-Related Failures by NSSS

Figure 5 shows the number of age-related failures of pressure instruments by NSSS. The left column of each pair shows the absolute number of age-related failures totaled by NSSS. The right column of each pair shows the total divided by the number of plants of that NSSS vendor. The figure shows that, on average, Westinghouse and Babcock and Wilcox plants report somewhat higher counts by NSSS than do Combustion Engineering or General Electric plants.

4.7 Failures Resulting from Sensing Line Problems

The age-related failures of the pressure, level, and flow instruments discussed to this point included failures of the sensors, transmitters, cables, controllers, indicators, and various subcomponents. They also include age-related effects on sensing lines and instrument isolation valves. Part of the reason for including the various groups of components is that frequently the licensees' problem evaluations were reported at a relatively high level. That is, the specific subcomponents repaired or replaced were not specified. The 1984 and newer LERs have improved event descriptions which lessen this problem.

Sensing line problems leading to pressure instrument problems were reported in 401 LERs. Personnel errors accounted for about 60% of the sensing line problems. Age-related problems accounted

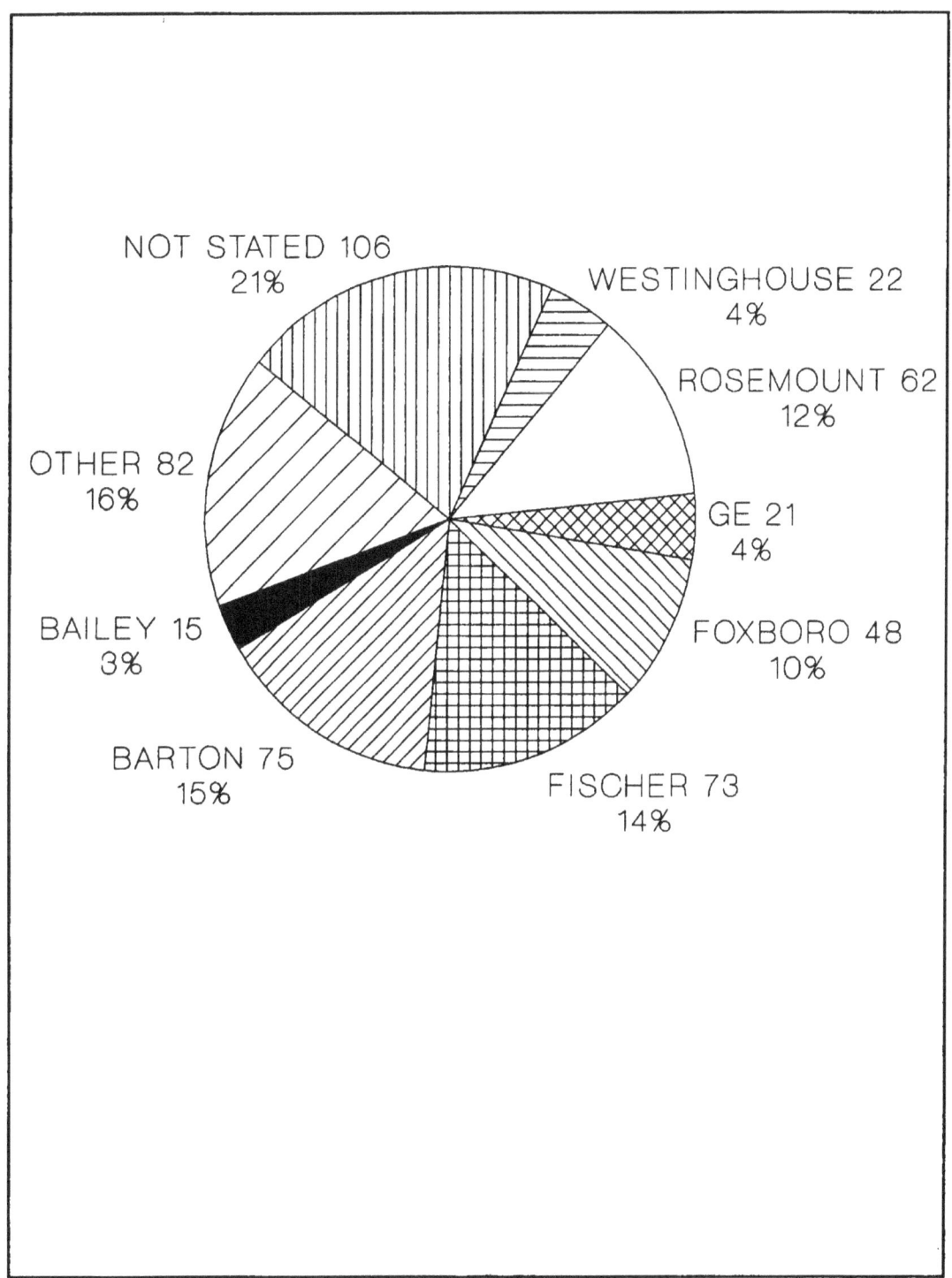

NOT STATED 106
21%

WESTINGHOUSE 22
4%

ROSEMOUNT 62
12%

OTHER 82
16%

GE 21
4%

BAILEY 15
3%

FOXBORO 48
10%

BARTON 75
15%

FISCHER 73
14%

Figure 4

A- 11

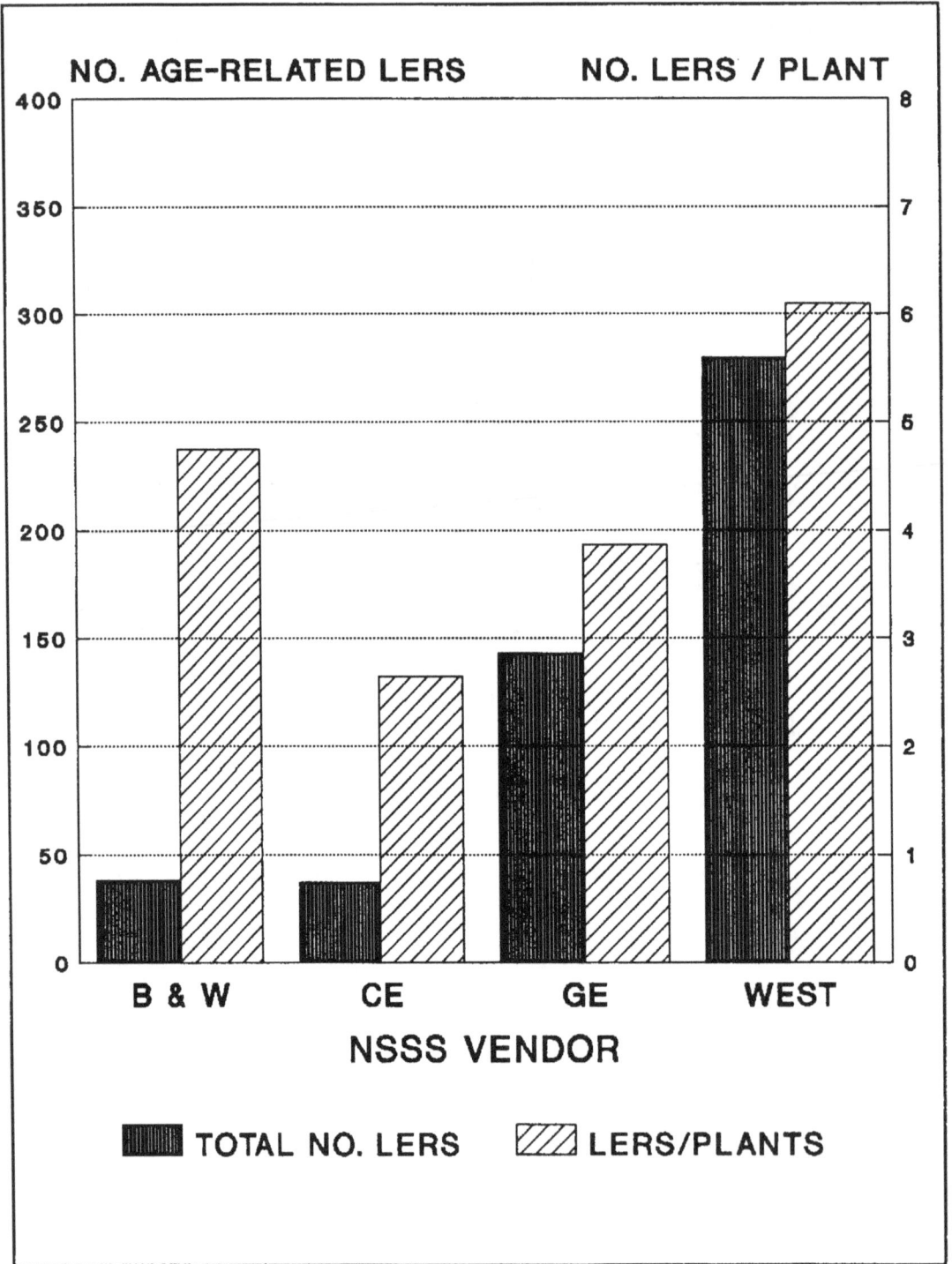

Figure 5

A-12

for about 27% of the sensing line problems. The most commonly occurring age-related problems are shown in Table 3. Problems with condensation, freezing, and crud buildup occurred most frequently. They accounted for approximately three-fourths of the age-related problems. Some of the less frequently occurring codes are probably do to problems related to either the pressure instruments themselves or to problems with miscellaneous types of subcomponents. Examples of these types of problems might include loose fittings, broken couplings, instrument calibration problems, etc.

Table 3 Causes of Age Related Sensing Line Problems

Category	Frequency
Water spray, condensation, flow blockage, or freezing	216
Worn, bent, broken, or damaged	35
Drift of calibration problem	27
Vibration or fatigue	4
Corrosion or erosion	5

4.8 Failures Resulting from Root Valve Manipulations

The effects of possibly undetected root valve manipulations or positioning have the potential to cause problem with pressure, level, or flow instrumentation. While age-related problems of root valves were included in counts of age-related instrumentation problems, they were not specifically called out.

The number of root valve problems is relatively small. Root valve problems were mentioned in 67 LERs. As might be expected, most of these problems were due to personnel errors - almost 70% of them. Age-related problems were reported in about 10%. Other, or unknown types of problems were mentioned in about 20% of the LERs. Binding or sticking problems were the most frequently occurring age-related problems in this small number of LERs. These codes were used four times.

5. CONCLUSIONS

A review of 1325 LERs reporting failures of pressure instrumentation over the 1980 through October 1988 time period was made. The purpose of the review was to identify age-related failures of these instruments. Age-related failures are defined as those failures which are due to effects of time, temperature, radiation, or other environment related effects. The results of the review identified 498 LERs reporting age-related failures. Pressure instrumentation problems were reported in about 4% of the total number of LERs written during this time period. Age-related effects were present in about 40% of these.

The review also identified 540 LERs where the pressure, level, or flow instrumentation problems were due to personnel errors. About 70% of the personnel errors occurred as design, maintenance, or testing problems.

Additionally, the study identified 342 LERs reporting pressure, level, or flow instrumentation problems which occurred for other, or unknown reasons. A specific cause of failure was not identified by the licensee in the majority of these events, nor could age-related causes of failures be reasonably determined based upon information in the LER.

LER information is a reasonable source of data to identify the types of pressure, level, or flow instrumentation problems apparent in the industry. LER data is not sufficient to determine exact numbers of pressure instrumentation problems because of plant dependent reporting requirements prior to 1984 and because of the change in LER reporting requirements in 1984 which reduced the number of LERs reporting these instrumentation problems. LER data is sufficient to characterize the major age-related problems affecting the instrumentation.

Age-related effects contributed to pressure instrumentation problems in 498 LERs. The most commonly reported age-related effects concerned drift or calibration problems (about 45% of the LERs); followed by water spray, condensation, flow blockage, or freezing (about 38% of the LERs); and worn, bent, broken, or damaged components (about 14% of the LERs).

Vendor codes associated with age-related failures were examined. Barton, Fischer, Rosemount, and Foxboro were the most frequently occurring vendor codes (about 10 to 15% each). No vendor code was given in about 20% of the cases. About 16% of the failures were attributed to an "other" category consisting of about 20 vendors.

Westinghouse and Babcock and Wilcox NSSS plants reported a higher number of pressure, level, or flow instrumentation problems per plant than did Combustion Engineering or General Electric plants. Westinghouse plants reported over twice the number as CE plants (which reported the fewest). Babcock and Wilcox plants reported about twice the number as CE plants. GE plants reported about one-third more than CE plants.

Sensing line problems contributed to about 400 LERs reporting pressure instrumentation problems, with about 60% due to personnel actions. Age-related problems (mostly freezing, condensation, or crud buildup) contributed to about 27% of the LERs.

Root valve problems associated with the pressure, level, or flow instrumentation were reported in a small number (67) LERs. Most of these were due to personnel errors.

APPENDIX B

FAILURE OF ROSEMOUNT MODEL 1153
AND 1154 TRANSMITTERS

NRC Information Notice No. 89-42

UNITED STATES
NUCLEAR REGULATORY COMMISSION
OFFICE OF NUCLEAR REACTOR REGULATION
WASHINGTON, D.C. 20555

April 21, 1989

NRC INFORMATION NOTICE NO. 89-42:   FAILURE OF ROSEMOUNT MODELS 1153 AND 1154
TRANSMITTERS

Addressees:

All holders of operating licenses or construction permits for nuclear power
reactors.

Purpose:

This information notice is being provided to alert addressees about recent
failures of Rosemount models 1153 and 1154 pressure and differential pressure
transmitters. It is expected that recipients will review the information for
applicability to their facilities and consider actions, as appropriate, to
avoid similar problems. However, suggestions contained in this information
notice do not constitute NRC requirements; therefore, no specific action or
written response is required.

Description of Circumstances:

During 1986 and 1987, five Rosemount model 1153 HD5PC differential pressure
transmitters malfunctioned at Northeast Utilities' (NU) Millstone Nuclear
Power Station, Unit 3. During power operation, the Millstone operators noted
that the signals from the Rosemount 1153 transmitters were deviating from
redundant channel signals and that the transmitters were indicating reduced
levels of process noise. The transmitters were declared out of service by
NU personnel, and the affected channels were placed in the tripped condition.
After attempts to calibrate the transmitters failed, NU returned the transmitters to Rosemount and informed them that the malfunctions had occurred
with transmitters of the same model and related serial numbers. Destructive
testing performed by Rosemount determined that the failures were caused by the
loss of oil from the transmitter's sealed sensing module. However, Rosemount
indicated that the failures appeared to be random and not related to any generic
problem with Rosemount 1153 pressure transmitters. NU submitted a 10 CFR Part 21
notification to the NRC on this issue on March 25, 1988, and provided additional
information on the failures via a letter dated April 13, 1989.

Discussion:

After additional evaluations by NU and Rosemount, Rosemount issued a letter
to its customers on December 12, 1988, regarding the potential malfunction
of models 1153 and 1154 pressure and differential pressure transmitters. The

Rosemount letter was supplemented with a letter dated February 7, 1989, to customers who had purchased transmitters from specific lots that were identified by Rosemount as being potentially defective. Rosemount issued a separate letter dated February 16, 1989, to customers who had purchased model 1153 and 1154 transmitters from lots that were not considered suspect. Rosemount indicated that transmitters from the suspect lots were susceptible to a loss of silicone oil from the transmitter sealed sensing module and to possible failure. According to Rosemount, as the oil leaks out of the sensing module the transmitter's performance gradually deteriorates and may eventually lead to a detectable failure.

Some of the symptoms that have been observed during operation and before failure include slow drift in either direction of about 1/4 percent or more per month, lack of response over the transmitter's full range, increase in the transmitter's time response, deviation from the normal signal fluctuations, decrease in the detectable noise level, deviation of signals from one channel compared with redundant channels, "one sided" signal noise, and slow response to a transient or inability to follow a transient. Some of the symptoms observed by NU personnel during calibration include the inability to respond over the transmitter's entire range, slow response to either increasing or decreasing hydraulic test pressure, and drift of greater than 1% from the previous calibration.

Although some of the defective transmitters have shown certain symptoms before their failure, it has been reported that in some cases the failure of a transmitter may not be detectable during operation. In addition, Rosemount now indicates that the potential for malfunction may not be limited to the specified manufacturing lots previously identified in the February 1989 letter.

It is important for addressees to determine whether any Rosemount models 1153 and 1154 pressure and differential pressure transmitters, regardless of their manufacturing date, are installed in their facilities and to take whatever actions are deemed necessary to ensure that any potential failures of these transmitters are identified. Although it may not be possible to detect the onset of failure in all instances, some transmitters have exhibited some of the aforementioned symptoms before failure. It is important for potential failure modes to be identified and that operators be prepared for handling potential malfunctions. In addition, careful examination of plant data, calibration records, and operating experience may yield clues that identify potentially defective transmitters. Addressees may wish to contact Rosemount for assistance in determining appropriate corrective actions whenever any of the aforementioned symptoms are observed or if failures are identified.

On April 13, 1989, the NRC staff met and discussed this matter with Rosemount and several industry groups. Rosemount has launched a program to identify the root cause of the loss of oil from the sensing module and to determine recommendations for its customers to address potentially defective transmitters.

No specific action or written response is required by this information notice. If you have any questions regarding this matter, please contact one of the technical contacts listed below or the Regional Administrator of the appropriate regional office.

Charles E. Rossi, Director
Division of Operational Events Assessment
Office of Nuclear Reactor Regulation

Technical Contacts: Kamal Naidu, NRR
(301) 492-0980

Jaime Guillen, NRR
(301) 492-1170

Attachment: List of Recently Issued NRC Information Notices

NRC FORM 335
(2-89)
NRCM 1102,
3201, 3202

U.S. NUCLEAR REGULATORY COMMISSION

BIBLIOGRAPHIC DATA SHEET

(See instructions on the reverse)

1. REPORT NUMBER
(Assigned by NRC. Add Vol., Supp., Rev., and Addendum Numbers, If any.)

NUREG/CR-5383

2. TITLE AND SUBTITLE

Effect of Aging on Response Time of Nuclear Plant Pressure Sensors

3. DATE REPORT PUBLISHED

MONTH	YEAR
June	1989

4. FIN OR GRANT NUMBER

D 2503

5. AUTHOR(S)

H. M. Hashemian, K. M. Petersen, R. E. Fain, J. J. Gingrich

6. TYPE OF REPORT

Technical

7. PERIOD COVERED *(Inclusive Dates)*

September 30, 1988 to
March 30, 1989

8. PERFORMING ORGANIZATION – NAME AND ADDRESS *(If NRC, provide Division, Office or Region, U.S. Nuclear Regulatory Commission, and mailing address; if contractor, provide name and mailing address.)*

Analysis and Measurement Services Corporation
AMS 9111 Cross Park Drive, NW.
Knoxville, TN 37923-4599

9. SPONSORING ORGANIZATION – NAME AND ADDRESS *(If NRC, type "Same as above"; if contractor, provide NRC Division, Office or Region, U.S. Nuclear Regulatory Commission, and mailing address.)*

Division of Engineering
Office of Nuclear Regulatory Research
U.S. Nuclear Regulatory Commission
Washington, DC 20555

10. SUPPLEMENTARY NOTES

11. ABSTRACT *(200 words or less)*

Research was initiated for the effects of normal aging on performance of pressure transmitters in nuclear power plants. This began with an experimental assessment of the methods used for response time testing of transmitters, followed by aging tests on representative transmitters.

The project included a search of the LER data base and a review of the Regulatory Guide 1.118 and the related standards. The conclusions are:

- Five reasonably effective methods are available for response time testing of pressure transmitters, two of which provide on-line measurement capability.

- The consequences of normal aging were calibration shifts and response time degradation, with the former being the more pronounced problem.

- The LER data base contains 1,325 cases of reported problems with pressure sensing systems over a nine year period. Potential age-related cases account for 38 percent of these problems.

- Regulatory Guide 1.118, IEEE Standard 338, and ISA Standard 67.06 can benefit from minor revisions to account for recent advances in performance testing technologies.

12. KEY WORDS/DESCRIPTORS *(List words or phrases that will assist researchers in locating the report.)*

Aging Degradation
Pressure Transmitters
Safety Related
Sensing Line
Response Time Testing

On-Line Testing
LER Data Base
IEEE Standard 338
Regulatory Guide 1.118

13. AVAILABILITY STATEMENT

Unlimited

14. SECURITY CLASSIFICATION

(This Page)

Unclassified

(This Report)

Unclassified

15. NUMBER OF PAGES

16. PRICE

NRC FORM 335 (2-89)